Evolution and Eugenics in American Literature and Culture, 1880–1940

Evolution and Eugenics in American Literature and Culture, 1880–1940

Essays on Ideological Conflict and Complicity

Edited and with an Introduction by

Lois A. Cuddy and
Claire M. Roche

Lewisburg
Bucknell University Press
London: Associated University Presses

Associated University Presses
2010 Eastpark Boulevard
Cranbury, NJ 08512

Associated University Presses
Unit 304, The Chandlery
50 Westminster Bridge Road
London SE1 7QY, England

Associated University Presses
P.O. Box 338, Port Credit
Mississauga, Ontario
Canada L5G 4L8

The paper used in this publication meets the requirements of the American National Standard for Permanence of Paper for Printed Library Materials Z39.48-1984.

Library of Congress Cataloging-in-Publication Data

Evolution and eugenics in American literature and culture, 1880–1940 : essays on ideological conflict and complicity / edited and with an introduction by Lois A. Cuddy and Claire M. Roche.
p. cm.
Includes bibliographical references and index.
ISBN 0-8387-5555-0 (alk. paper)
1. American fiction—20th century—History and criticism. 2. Evolution (Biology) in literature. 3. American fiction—19th century—History and criticism. 4. Darwin, Charles, 1809–1882—Influence. 5. Literature and science—United States. 6. Evolution (Biology)—United States. 7. Eugenics—United States. 8. Eugenics in literature. 9. Race in literature. I. Cuddy, Lois A. II. Roche, Claire M.

PS374.E88E95 2003
810.9′356—dc21 2003005341

PRINTED IN THE UNITED STATES OF AMERICA

We dedicate this volume
to Eamon Kenton Connolly, Miles Thomas Treichel,
Neve Thalia Treichel,
and to future grandchildren

to Gregory Rust McNab

and

to Mickey Hazard, whose ongoing support
makes all things possible.

Contents

Introduction: Ideological Background and Literary Implications

When Charles Darwin published *The Origin of Species* in 1859, he presented a theory of descent from a common origin that challenged Western assumptions of theology, time, and human existence. Though he centered his observations and collected evidence on organic life, the flora and the fauna on Earth, the implications for the origin and development of human life and for humanity's view of itself were awesome. Most people in the twenty-first century take for granted the overwhelming evidence in favor of the validity of evolution, for the advances in sciences have verified that life on Earth is many millions of years old and that primates and human beings share most of their DNA. However, in Darwin's time, his theory of origins and "descent with modification through natural selection" (Darwin 1985, 435) shattered religious and societal complacency even while it excited researchers across the sciences. After 1859, Western ways of seeing the world and the nature of life were never the same. In 1894, Benjamin Kidd in his second American edition of *Social Evolution* summarized the extent of that influence: "One of the most remarkable epochs in the history of human thought is that through which we have passed in the last half of the nineteenth century. The revolution which began with the application of the doctrines of evolutionary science, and which received its first great impetus with the publication of Darwin's *Origin of Species*, has gradually extended in scope until it has affected the entire intellectual life of our Western civilisation" (vii). Darwin's concepts of descent from animal species, of scientific determinism, the struggle for life, adaptation, and progress made their way into all aspects of intellectual production and influenced the development of eugenics and social sciences as part of this new evolutionary paradigm of life. It makes sense, then, that literature—in representing, challenging, and critiquing culture—would appropriate and aesthetically transform this ubiquitous theory. By addressing the extent and depth of that influence in

American literature, the present volume attempts to add new voices and perspectives to this discussion.

The intensity of debate over evolutionary theory and its implications has been remarkable and enduring. Darwin's organic theory of descent with modification, later called "evolution" by Herbert Spencer,[1] resonated throughout the Western world when Darwin concluded that "probably all the organic beings that have ever lived on this earth have descended from some one primordial form" (Darwin 1985, 455).[2] And with that startling statement, man's[3] organic relation to all of life linked him to the animal kingdom. Thus, for many intellectuals, a literal reading of the King James translation of Genesis took its place in the pantheon of questionable and discarded truths through the ages. With the proliferation of Darwin's (and other scientists') collected evidence, gone forever was the certainty that God created each discrete form of life fully developed and with its purpose defined at one moment in time. No longer could the earth have been created a mere six thousand years ago (Burrow 1985, 20), and no longer were God and man in his likeness at the center of life with the absolute power to create and name all that exists.[4]

While many people held to traditional doctrine and rejected the new science, members of the scientific establishment like T. H. Huxley denounced the validity of religion in the interests of evolutionism. Other notable thinkers like Herbert Spencer sought to accommodate evolutionary science with the spiritual.[5] William James, one of the most respected and influential philosophers of his time, concluded in a lecture in 1895 that "possibilities, not finished facts, are the realities with which we have actively to deal" and the most we can hope for is to believe "that the possibility exists" (quoted in Pizer 1972, 29). Western belief and theology were no longer monolithic as science gradually replaced biblical absolutes as the arbiter of truth, right and wrong, and the questionable purpose of life. The results of these uncertainties are evident in both European and American literatures.

With Darwin's publication of *The Descent of Man* in 1871, the possibility that man evolved and was modified through natural selection over millions of years could not be ignored. Scientists now turned their attention to *how* this happened and to *how* their own sciences could participate in the debate over this slow and inevitable process. In *The Descent of Man*, Darwin applied some of his most daring and challenging concepts to human life, and these ideas became an integral part of education, the emerging social sciences, and popular culture such as newspapers and magazines. Attitudes and values began to change in the United States as notions of struggle and competition in the survival of the fittest became a way of perceiving life and human relationships. For example, adaptation to environmental and hereditary forces challenged free will and the efficacy of prayer; scientific determinism supplemented (or even supplanted) God as the reason for all things; and people in the wealthy and educated classes applied the paradigm of evolution to history, literature, imperialism, racial and class policies, gender issues and birth control, eugenic ideology, and to the belief in individual as well as phylogenic progress (or regression). The power of evolutionary theory was so profound that in 1878 the prestigious *Encyclopaedia Britannica* dedicated twenty-eight double-column pages to the definitions and history (or evolution) of the subject. The discussion of "Christianity" was assigned only fourteen pages in that edition. By the end of the nineteenth century, evolution, with its social applications, became a way of thinking about the world as it permeated all aspects of personal, social, and political life. Darwin's ideas, in both accurate and distorted forms, became part of popular culture.

Eugenics in the United States

The American fascination with progress is not unique to Darwinian theory or to the eugenics movement. Progress has long been a concept and a paradigm at the center of our culture, and scientific developments confirmed its possibilities as never before. Our history as a nation is based on the assumption that progress is necessary, natural, and desirable—that it is by its very nature good and beneficial. While that may be an accurate assessment, what has been done in the name of improvement is often more troublesome.

One of the results of Darwin's work was a drive to advance the human race through a pseudo-science called eugenics—the "science" based on heredity designed to improve the human race by selective breeding of the "fittest" families. By the early twentieth century, many Anglo-Americans were focusing on such improvement as a measure of personal as well as social and national progress. In 1909, for example, Reverend William Inge (a neighbor to Francis Galton, the founder of eugenics) argued that "Progress, for the moralist and for the biologist alike, means improvement in the people themselves and not in their conditions"

(1). Like so many other English and American scientists and thinkers, Inge believed that if biologists could advance human breeding by giving evolution a helping hand, the conditions in which people lived would inevitably be meliorated. Eugenics was hailed as the method to insure that progress in the Anglo-Saxon race and, in fact, we might safely invoke "progress" as both the trope and the obsession of the eugenics movement in England and in the United States. This obsession spread through Europe and eventually to Japan and Latin America (Kevles 1986, 63); it reached its peak, as we know, with the pursuit of the master race and the final solution by Adolf Hitler and the Nazi regime. Yet, though eugenics as an organized political force came virtually to an end with the Nazis, many ideas and social policies initiated by the eugenics movement remain ingrained in attitudes about race, class, poverty, sexuality, and responsibility in the United States to the present time.

We can look to 1865 as a defining moment in the development of eugenics as an ideology and a "science," which began as something of an experiment in human heredity and statistical methods. In the year prior to the publication of Gregor Mendel's significant but long-overlooked research in genetics, Francis Galton published a two-part article in *Macmillan's Magazine* in which he outlined his eugenic ideas, expressed his belief in the progress promised by evolution, and wondered whether human beings could intervene in their own evolution and natural selection to hasten the advancement due them. Although he didn't actually coin the phrase "eugenics" until 1883, Galton arrived at the name by using a word with a "Greek root meaning 'good in birth' or 'noble in heredity' " (Kevles 1986, xiv). Eugenics meant the "science" of improving human stock by giving "the more suitable races or strains of blood a better chance of prevailing speedily over the less suitable" (xiv). Since hereditary research in the nineteenth century was a new field of inquiry and a highly inexact one, experiments in heredity conducted by Galton and his followers at his laboratory in England often resulted in flawed and questionable results for which Galton was nevertheless rewarded and even knighted (57).

In 1907, while Galton was actively lecturing on eugenics in England, Charles Davenport was conducting research in the United States on the Mendelian inheritance of hair and skin color and played a significant role in furthering the eugenics agenda in America (57). Davenport, a zoologist from a prominent New England family, had spearheaded the establishment of a research center, the Eugenics Record Office at Cold Spring Harbor, New York. Funded by the Carnegie Institute, this center was a recipient of one of the first research endowments established in this country. At that site, Davenport collected thousands of family pedigrees on which he based his research (45). He was also one of the principal scientists concerned with studying "feeblemindedness" and "negative eugenics," which called for preventing the reproduction of the genetically "defective" members of society (45). Moreover, Davenport worked to improve human germ plasm, the substance believed to hold genetic secrets and material, for he believed that behaviors, like diseases or physical features, were hereditary. He was one of the busiest and most productive—and one of the most racist, sexist, and ethnocentric—eugenicists in Europe and the United States. He argued that feeblemindedness, pauperism, and prostitution (the latter practice often associated with Jewish women in the eugenicists' attempt to limit immigration of that group to America) could be explained as having hereditary connections to "primitive genes" (Rosenberg 1997, 91). His agenda becomes clear in a letter to a friend in which he wrote that "our ancestors drove Baptists from Massachusetts Bay into Rhode Island but we have no place to drive the Jews to. Also they burned the witches but it seems to be against the mores to burn any considerable part of our population" (93–95).

As absurd as Davenport's views now seem, he and his followers were influential in passing eugenics laws and in helping to popularize the American eugenics movement, which peaked in the 1920s and 1930s. As an example of the widespread cultural acceptance of eugenic ideas, "Fittest Family" contests were held in the "human stock" sections of state and county fairs in the 1920s (Kevles 1986, 61–62). By 1923, the American Eugenics Society was firmly established, and with the help of Davenport, chapters and eugenic education societies were active in many states (59). Davenport didn't propel the movement alone, however. All over the country intellectually and socially prominent men and women joined clubs, hosted lectures, contributed articles, and lobbied for legislation to support their eugenic agendas. Lawmakers in Washington, D.C., were led by people like Congressman Albert Johnson, chair of the House Committee on Immigration and Naturalization, who commissioned a study by Harry H. Laughlin, Davenport's "right-hand man at Cold Spring Harbor" (102). In his report to the House Committee, Laughlin concluded "that the recent immigrants were biologically inferior and that they jeopardized the blood of the nation" (103). Despite his distorted "evidence," an "authority" like Dr. Laughlin had enormous influence by his "endorsement of the permanent immigration restriction bill" (103), which was subsequently passed. In this atmosphere of national fear, the 1920s also saw legislation legalizing involuntary sterilization in Virginia, California, and twenty-two other states. In the Supreme Court decision in *Buck v. Bell* in 1927, the Court upheld a Virginia statute that allowed for the involuntary sterilization of Carrie Buck, a young woman institutionalized at the Virginia Colony for Epileptics and the Feebleminded in Lynchburg (110–11). Tens of thousands of Americans would be sterilized subsequent to this court decision, and although in greatly reduced numbers, the practice continued in America beyond the 1930s. Only in the 1970s, when the ACLU brought a suit against the Lynchburg Colony, did that institution end its use of involuntary sterilization.

The pattern of social and cultural changes taking place after the Civil War created the opportunity for eugenicists to go public with their ideas and agenda. Specifically, the combination of urbanization, industrialization, and increasing secularization taking place between the Civil War and the Great Depression created the circumstances under which eugenics could prosper. America's love affair with science and proofs and facticity began here, in no small part due to the work of Charles Darwin, whose attitudes about "the imbecile, the maimed, and other useless members of society" (1981, 1:103) surely had an impact on the ideology that resulted in the sterilization and institutionalization of many people with physical and developmental disorders. Technology began to take on a life of its own while, ironically, greater numbers of laborers were required to produce the labor-saving advances entering the marketplace. More than ever before in the United States, urbanization brought more people into worse conditions under closer scrutiny and in closer proximity to the upper and middle classes. To illustrate this point, it is worth noting that in 1860 there were sixteen cities with populations over fifty thousand. By 1910, more than eighty cities had populations in excess of fifty thousand (Budd 1995, 29). This is accounted for, in part, by the fact that twenty-five million people arrived in America between 1860 and 1920. We can get a sense of what this number means if we consider, for example, that currently the states of Arkansas, Alaska, Delaware, Montana, North Dakota, South Dakota, Rhode Island, Vermont, Wyoming, and the District of Columbia have fewer than twenty-five million residents combined (Bureau of the Census 2000). The social and economic implications of these numbers had a direct influence in the passing of the Johnson Act in 1924. With the advent of nationwide transportation and the technology of photography, people all over the United States could get a glimpse at urban life even if they didn't have to live it. In 1890, Jacob Riis published *How the Other Half Lives*, his contribution to social reform for the urban poor. Using photography to capture America's huddled masses living in squalor on the lower east side of Manhattan, Riis introduced middle class and rural America to what might have been their deepest anxieties about the consequences of immigration and poverty. Riis's photographs, like many of the naturalistic novels, were populated by poor people to whom eugenicists would likely apply the label "feebleminded."

Thomas Shapiro, in his 1985 book, *Population Control Politics: Women, Sterilization and Reproductive Choice*, argues that the eugenics movement grew in America because "ethnocentrism, a fear of status revolution, and a concrete political threat from America's working class provided a fertile ground for it" (38). He also argues that "pseudo-sciences like eugenics must be understood as a social phenomenon" (40). Certainly, no eugenicist working in America in the period under consideration was able to prove that eroticism was genetic, as Davenport claimed, or that human germ plasm could be manipulated to rid people of feeblemindedness. Yet, even reputable and well-known scientists, such as Princeton biologist Edward Grant Conklin, were unable to counter the popularity and widespread acceptance of eugenics as science (42). But although eugenicists really had no scientific proof on which to base their conclusions, they could, however, rely on threats to social cohesion and stability, threats posed by an increasingly restless and angry working class.

The threats gained resonance in a historical moment in which Americans became, more than ever before, fascinated with ideas about purity and authenticity and about conscious behavior and unconscious drives. David Shi, in *Facing Facts: Realism in American Thought and Culture, 1850–1920*, argues that Freud's ideas, and distortions of them, took hold fairly quickly in America, and that by 1916 there were approximately five hundred psychoanalysts in New York City. Shi quotes Susan Glaspell who remembered that "you could not go out to buy a bun without hearing of someone's complex" (282). Americans were captivated by "the notion of unruly and unpredictable sexual desires and aggressive instincts at the bottom of the psyche" (282). Reinforced by evolution's requirement for reproduction of the best human specimens and by Victorian concepts of morality, this obsession with the pathology of desires and instincts, typically associated with homosexuality, poverty, and the working class members of American society, was accompanied by a sense of revulsion. It was this response on which eugenicists relied in furthering their agendas.

With post-Civil War industrial and technological development came an increasingly ambivalent relationship to the working class on the part of the middle and upper classes in America. Advances in science and engineering, for example, led to increases in manufacturing productivity. As productivity and revenue increased, however, so too did a reliance on working class labor (although often in fewer numbers or in ways that required either no skill or much more technical skill) to keep it all going. While the eugenics movement worked to increase the birthrate of "desirable" parents, they encouraged restraint and sterilization for those less desirable parents who made up much of the labor force. One of the paradoxical consequences of this movement to restrict reproduction of the "unfit," then, was that the supporters of eugenics were caught between the ideal of eugenic progress through the reproduction of the "fittest" members of society and the reality of a necessary labor force populated by the "undesirables."

The sciences that spurred industrialization also hastened the trend toward secularization in America. In the late-nineteenth century, "sciences were less abstruse than today, and their leaders wrote for the weekly and monthly magazines" (Shi 1995, 81). Scientists were, as they are now, voices of cultural authority, especially when claiming that they can produce the real, the true, the authentic, the pure. Simply put, scientists heralded progress, and progress was the goal and myth of American experience. Genetic advances were as desirable as artistic, commercial, or territorial progress: "the mood which characterized the growing use of hereditarian ideas in the middle third of the [nineteenth] century was one of confidence that man's most fundamental attributes could and should be manipulated" (Rosenberg 1997, 36). In an increasingly secular culture, it became possible to make the argument that man, not God alone, could intervene in the process of creation. The mood to which Rosenberg refers resulted from secularization, in part, and made it possible for eugenics to coalesce into a social and pseudo-scientific movement in America.

American popular fascination with science has historically taken many forms. One aspect of scientific influence was the growth in popularity of "realism" in American literature, art, and architecture. Even writers who did not work in realism were often intrigued by, and therefore often made use of, science in their work. Rosenberg, author of *No Other Gods: On Science and American Social Thought*, argues that in the nineteenth century, largely as a result of Darwin's work, science became part of the culturally "accepted and acceptable knowledge" and that this knowledge was very much a part of, indeed inseparable from, the social order in America. As a result, he says, "to reject an idea endorsed by men of learning was to reject, at least partially, order and stability in society" (1997, 3). Thus, when science in the forms of Darwinism and eugenics became part of the intellectual and social fabric of American society, these sciences emerged in works of realism and naturalism as well as utopian and modernist drama and fiction and influenced both classes and individuals. The essays in the present volume also reveal that American literature—whether expressing ambivalence, resistance, or complicity—was shaped by the scientific theories that generated support for the social ideologies and political policies of the age. Little wonder that authors writing as realists, naturalists, modernists, or socialists would be seduced, consciously or unconsciously, by the evolutionary and eugenic ideologies that place the eminent authors themselves on the highest level of human development.

Hierarchy and Social Values

Western belief in evolution's long history of gradual modification and progress justified long-ingrained attitudes of superiority in the white, educated, and well-to-do Anglo-Americans. Darwin's texts, both their evidence and language, became the scientific authority for that Anglo-Saxon preeminence. For example, his contrast of undeveloped "savages" and "barbarians" with the "higher" and more intellectually evolved "civilised" [*sic*] man (woman had not evolved to the supreme level of man in Darwin's scheme) reveals the white, male, elitist propensities and prejudices that were reproduced in all aspects of American society. These attitudes of superiority and privilege, reinforced by eugenicists and social Darwinists,[6] further inflamed the prejudices in American society (Kevles 1986; Shipman 1994) and found their way into literature (Cuddy 2000). Widespread anti-Semitism was certainly not new to Western societies (Almog 1988), but it was further fueled by scientific "evidence" and a rhetoric of bigotry that assigned Jews to the primitive and lower stage of development (Pizer 1972, 23) along with Africans (Achebe 1989) and anyone else different from people with a northern European heritage. Presumably white Americans now had scientific support for the long-held belief that African Americans were comparable to the ape species but surely not to the Christian white man's "higher" intellect. One need no more incontrovertible evidence than the color of skin and inferior character—qualities that were somehow assigned to blacks, Jews, women, and the poor—to recognize the closer affinity of these groups with the animal kingdom from which they apparently evolved more slowly than well-to-do white men. Although nationalistic prejudice and assumptions of the superior man's rights over inferiors are older than ancient historical texts, the difference is that in the later nineteenth century both hierarchy and bigotry were not only culturally approved but also encouraged by science. It is hardly a coincidence that such scientific research and publications were produced by white men from the privileged class. American literature and popular culture reflected these ideas, typically without self-analysis.

In this regard, it is important to note Darwin's rhetoric in *The Descent of Man*, for his language represents and further justifies white man's belief in his own superiority. The scale of evolutionary development is from the "low" to the "high," and these are clearly value words. Darwin correlates "the lowest savages" (e.g., Africans and islanders who are different from northern Europeans) with "the most highly organised ape" and points out that savages cannot think abstractly (Darwin 1981, 1:34). He quotes "the maxim of the Spaniard, 'Never, never trust an Indian'" (1:95) in order to illustrate the savagery and ethical defects of the uncivilized Indian without recognizing the defects of the (white) Spaniards who murdered and enslaved those very Indians in the New World. His text is replete with such comments.

Darwin ended *The Origin of Species* with an optimistic view about the future of humanity: "And as natural selection works solely by and for the good of each being all corporeal and mental endowments will tend to progress towards perfection" (1985, 459). In *The Descent of Man* he still believed that "man is the codescendant with other mammals of a common progenitor" (1981, 2:386) and therefore is descended from the primates who inhabited the African continent (1:199). And while "we are descended from "wild animals," "barbarians," and the "savage" (2:404–5), civilized man has made enormous progress in developing beyond our progenitors. Darwin also pays tribute to the development of white Americans because of the quality of the Europeans who emigrated to the United States, but he concludes that the highest level has been achieved by the light-skinned, Anglo-Saxon Europeans: "The western nations of Europe . . . stand at the summit of civilisation" (1:178). However, not all "descendants" or western Europeans had made equal progress toward "civilisation."

Darwin gradually modified his assumption of progress in the human race. After examining the studies of peoples around the world by anthropologists and other scientists, Darwin began to think in more relative terms. For example, Darwin accepts the biases of his cousin Frances Galton, the father of eugenics, whom Darwin quotes with admiration in *The Descent of Man*, and concludes that the Irish have shown a "downward tendency": While the moral and spiritual Scot—the "frugal, foreseeing, self-respecting, ambitious . . . [Scot who is] sagacious and disciplined in his intelligence"—marries late and produces few offspring, "The careless, squalid, unaspiring Irishman multiplies like rabbits" (1:174). Darwin and Galton warn that if this condition continues, "it would be the inferior and less favoured race that had prevailed—and prevailed by virtue not of its good qualities but of its faults" (1:174). The poor, like some nationalities and races, for instance, were deemed unfit to reproduce and populate the earth, and the wealthy and educated classes were urged to have more children in order to retain power through numbers. Thus, language like "downward," "inferior," and "less-favoured" assigns selected peoples to the realm of the unfit and illustrates only a few of the many value judgments and prejudices in the research and "objective observations" of nineteenth and twentieth-century science. The influences of these attitudes in Western society, in education and the culture of the educated, and in the authors who perpetuated these ideas about inferior races and people in poverty cannot be ignored. It is no coincidence, for instance, that Stephen Crane's Maggie is from an Irish family of drunks and profligates.

Even the gendered language assigning all measures and standards of development to males, to "man," "he," and so on, hierarchizes male and female. And we cannot fall back on the complacent assumption of the "universal" male diction because in the relatively few instances when Darwin wants to make a point about women and their inferior development, he makes the distinction between man and woman. Clearly he is concerned with "man" and the standard "he" has achieved both physically and mentally. Darwin's belief in the superiority of man over woman in terms of not only physical size, strength, courage, and energy but also "intellectual vigour and power of invention" (1981, 2:382) assigns a lower place and limited progress to woman. His point of view, then, is that females have not progressed very far since even the most admirable qualities held by women ("intuition," "rapid perception," and "imitation") are "characteristic of the lower races" (2:326–27). The cultural acceptance of man's "higher eminence" and intellectual superiority of "thought, reason, or imagination" (2:327) over that of woman is given credence by Darwin's conclusions and pointed rhetoric. School textbooks reflected these attitudes (Cuddy 2000), and the literature of the late nineteenth and twentieth centuries further reinforced the sexism that has historically driven gender relations in Western societies.

Many women authors were complicit in perpetuating the views of female developmental inadequacy. For example, Kate Chopin's Edna in *The Awakening*, like Ibsen's Hedda Gabler, could not face the realities of herself or the difficulties of living and committed suicide in the ultimate statement on her unfitness to survive. Unlike men who have the strength to confront their problems and endure, women apparently had not progressed since the beginning of Western history. Though we now recognize that Charlotte Perkins Gilman wrote "The Yellow Wall-Paper" as a probable indictment against the medical treatment and family attitudes toward women who were suffering from legitimate ailments, the protagonist in that story was perceived as weak and demented in her time. Moreover, any woman who dares to think she is equal to man, who behaves like a man in her misguided freedom to choose what she desires, must pay a high price, as she has done throughout history. But after the first wave of the feminist movement in the nineteenth century, one might think that attitudes toward women would be different. Yet, Lillie Devereux Blake's *Fettered for Life* in 1874 shows the devastating consequences of male dominance, female submission, and the fitness to survive, according to cultural proscriptions. Grace Farrell (1996) shows us how Blake resists such conclusions and challenges society's beliefs that men in power have the right to destroy vulnerable women and that, as the inferior part of the human species, woman must be kept in a subordinate position in order to insure her protection and worthiness to reproduce and maintain racial strength and progress. Thus, female sexual and social freedom must be denied and even punished by society, as we also see in Edith Wharton's work and for different reasons in Hemingway. Neither Lily in *The House of Mirth* nor Catherine in *A Farewell to Arms* had the qualities that their societies (or religious doctrines of the time) deemed "fittest" for survival into the next generation. They must be killed off before reproducing, while the male protagonists continue to live their lives exactly as they choose and presumably continue to enjoy the admiration of others for doing so. Thus, social hierarchy, while surely not originating with Darwin, was given added credence by his scientific conclusions, and American authors—for the demands of a publication market and/or their own beliefs—participated in the replication of these attitudes, as the essays in this volume illustrate.

The rebellion against Romantic catharsis and metaphysics in the interest of realism, naturalism, and modernism in subject matter and aesthetic form and style seems to suggest that literature has the cultural independence and power to transform society in its own revolutionary image. However, such a conclusion requires examination. Just as Darwin's work did not emerge from a vacuum but rather was influenced by his grandfather's theory of evolution and adaptation in *Zoonomia* (Irvine 1955, 85), by "[t]he Malthusian principle [that] reflects a deterministic, quantitative, Newtonian mechanistic conceptualization of the world" (Schweber 1985, 35), and by the history of Western thought, so literature also does not spring from a cultural void. American writers were not necessarily ahead of their time, as humanists have wanted to believe; they simply reflected and sometimes challenged their culture's values, and the evolutionary theories and prejudices that permeated the literature reveal authorial complicity in some of the most exciting and destructive ideas in history. As Lars Ahnebrink notes, "Darwin's interpretation of the mechanism of the universe, the positivism of Comte, and Marx's socialism were adopted by the naturalistic school. Mill, Spencer, Buckle, Huxley, Haeckel, and other scholars substantially influenced the new generation. The new scientific, philosophical, and social ideas were merged in the naturalistic novel and gave to it its specific character" (Ahnebrink 1961, 22). Authors absorbed the ideas circulating in society and appropriated aspects of those theories that had transformed the ways we conceptualize human life and history.

Consequently, literary protagonists were no longer concerned with the Romantic search for spiritual and metaphysical answers couched in metaphor; rather, realism led to naturalism, whose authors set their characters in the morass of materialism and chance. Survival in literature meant not only Darwinian reproduction but also, and more significantly, economic and physical survival. The Romantic hero's freedom was replaced by scientific determinism, while evolution and eugenics informed the thinking, themes, and aesthetic structures of authors from William Dean Howells, Mark Twain, Frank Norris, Stephen Crane, W. E. B. Du Bois, and Pauline Hopkins, to Jack London and Edgar Rice Burroughs[7] (Taliaferro 1999), to T. S. Eliot, William Faulkner, Nella Larson, Thornton Wilder, Tillie Olsen, John Steinbeck, and many others. These authors and those represented in this volume were all well versed in evolutionary theory, and they in turn influenced later authors throughout the twentieth century.

Realism and Naturalism

Human Nature, Class, and the Survival of the Fittest

While Darwin began his work by studying physical characteristics and changes in a wide variety of plants and animals, in *The Descent of Man* in 1871 he discusses the development and importance of habits, instincts, moral and ethical behavior, social community, and qualities like love, conscience, kindness, and benevolence for survival of the human species. The notion of the survival of the fittest in relation to inherited traits and response to environmental factors became fertile conceptual ground for literary analysis of human nature and society. The literature of realism and naturalism borrowed Darwin's (and Spencer's) construct of the survival of the fittest, critiqued the definitions and implications of "fitness," measured the "civilized" qualities in human nature against life as authors observed it, and considered how values and relationships are altered when human beings are required to struggle alone for survival in a mechanistic, indifferent universe devoid of a benevolent God's aid or justice. The image of life reflected in the Darwinian mirror was not a comforting one.

Realism's fidelity to the details of contemporary, everyday life almost required that those authors document class structure, define the "fittest" within class conflicts, and reveal the injustices perpetrated on the poor by the rich. Authors like Hamlin Garland saw "nobility in the commonplace" (Pizer 1998, 5), for he "associated realism with 'a democratization of literature,' as Pellew put it in 1891" (5); however, William Dean Howells, for decades the de facto Dean of American letters, was concerned with "the emerging debate over social justice in America: the fear engendered by an empowered lower class, the anger stimulated by the obstacle to reform the present in an entrenched and obdurate upper class" (4). As Edwards reveals in his essay in this volume, Howells also struggled with ambivalent attitudes about race, as so many authors did. Despite the impulse for social reform in this literature, then, there is "a class bias that runs powerfully through much of the debate over realism" (4), for "as late as 1893 Maurice Thompson could note that 'we find that in fiction and poetry we are hobnobbing with persons with whom we could not in real life bear a moment's interview'" (5).

Thus, the subtle, and often patronizing, superiority in much of realism served, like Darwinism, to reinforce "higher" and "lower" class attitudes and to judge the worthiness of the protagonists in society. What Pizer says about sexuality can also be applied to attitudes about the poverty class in this literature: "Whatever the commitment by such defenders of realism as Howells and Garland to the 'modern' ideal of scientific objectivity, they shared with almost all critics of the day an acceptance of a traditional ethical dualism in which man's sexuality [class] was associated negatively with his animal past and was therefore a retrogressive element when found either in society or fiction" (Pizer 1998, 9). From the time of Hamlin Garland's "The Return of a Private," which is set on poor farms after the Civil War, to Crane's Maggie on the streets of New York, to the work on race by Charles Chesnutt and Pauline Hopkins, to upper class and anti-Semitic attitudes in Edith Wharton's and F. Scott Fitzgerald's (and many other authors') work, the representations of class and race and the implications for society and the individual are ever-present in this literature.

The confrontations between inherited wealth—with the putatively superior qualities inherent in that birth—and the newly-emerging middle class, and between the middle and lower classes permeate the writing of realist writers like Henry James and William Dean Howells. In James's *Daisy Miller* in 1878, for example, Winterbourne is as cold and inflexible as his aristocratic family requires as he helps to maintain class purity and superiority; while Daisy—the lovely young woman with Democratic principles of equality, with new money and the naive belief that clothes and honesty could challenge and match the entrenched Old World society—must die. Clearly, Winterbourne survives, for he represents the ruling social class and therefore is the fittest to adapt to the cruelties and diseases of his world. In *The Rise of Silas Lapham* in 1885—Howells's critique on capitalism, the god of money, and values—the battle between rich and middle class is resolved by the fittest members of the two families uniting in marriage. This is a variation on the racial amalgamation theory of the time, which believed that intermarriage would raise the "lower" (black) race up, for in this novel it is the expectation that the future generation will enjoy the best of both parents' heritage since both sides are white. At the end, the character, undeveloped intelligence, and ethical strength of the middle class merge with the intellect, beauty, and culture of the upper class as both Penelope Lapham and Tom Corey illustrate the adaptation that will mark the fittest of the species for survival in the economic and industrial new world order. As seen here, literary realism established a hierarchy of values in which the middle class might have more admirable values in terms of conscience, honesty, industry, common sense, and loyalty to friends and family over established society; however, members of the higher class survive and even flourish economically and socially on their old money or, in the case of the Coreys, on the son's ability to adapt to and participate in the new industrial society.

Mark Twain's work also shows the influences of ideological and social changes suggested in *The Descent of Man*.[8] In his *Adventures of Huckleberry Finn* (1884), a novel in which realism of style and social critique unite with the romantic hero, Twain depicts many classes on various levels of the evolutionary ladder. Huck's Pap is described as primitive in his physical appearance, violent behavior, lack of education, and weakness for alcohol. He has developed no "civilized" qualities. On the other hand, Jim, the slave, is an example of moral development but primitive thinking based on superstition and myth. Like the savages described by Darwin, Jim and the uneducated whites in this novel function without logic or rationality and reject information that does not fit their beliefs. To readers today, Jim is one of the most admirable, generous, and loving characters in American fiction. Yet, in the context of Darwinian theory, which permeated cultural attitudes in Twain's time, Jim is diminished. Like some of the higher and domestic animals in Darwin's scheme, Jim behaves with loyalty, love, strong social connection, and deference to higher authority, but without the capacity for abstract thinking. Either because of racial inferiority or his slave status, or both, he clearly has not evolved to the higher intellectual category of whites and will never progress to the level of the Judge, Aunt Polly, or even of Huck. Jim is an evolved human being in certain Darwinian aspects: he is a moral, kind, and gentle man with a developed conscience, and he is "fit" to "survive" through his children by passing his traits on to future generations; however, Twain (like Darwin) suggests that Jim's race has not progressed intellectually to the level of the whites who may be uneducated and even morally deficient, but they often reveal a shrewdness and adaptability that enable them to survive despite their weaknesses in character and conscience.

The principle of the survival of the fittest works in different ways in this novel. With the Grangerfords and Shepherdsons, for instance, the instinct to preserve their own at the expense of enemy tribes almost kills off the two families; their genetic survival depends on the marriage of the two young people who elope before the slaughter begins and thereby preserve the best of each family's inherited traits. Huck, the smart, clever, uneducated but adaptable protagonist and hero—who was initially ostracized and treated with contempt because of his poverty, background, appearance, and his family history—ultimately rejects society, its ignorance, and its hypocritical values and "light[s] out for the Territory" (Twain 1962, 366) in order to live apart from a society for which *he* now has lost respect. Thus, Twain joins the debate about whether progress is a result of inheritance or environment, but he offers no definite conclusion in this novel. We only know that with very little education, Huck's thinking progresses beyond Pap's and Jim's, and we also realize that only a white man like the Judge is superior in class, intellect, and ethics in this fictional world.

For authors like Mark Twain, the potential romantic freedom of characters like Huck Finn and Jim is dissipated by disturbing questions about scientific determinism. We see this conflict particularly in *Pudd'nhead Wilson* in 1894 in which opposing forces of nature and nurture are played out. When the light-skinned, mulatto baby is switched at birth with the "white" infant, Twain presumably offers the biracial baby an opportunity to surpass his biology and the cultural definitions of race. The "black" child is given a chance to become a noble and admirable man because his environment now treats him with all the dignity and privilege accorded a rich, "white" child. Yet, the text supports the influence of eugenics—the principle of human breeding and inheritance as the determining factor in a man's character and success. In fact, Twain read Francis Galton's 1892 book on fingerprints before he wrote *Pudd'nhead Wilson*, and "topics of slavery and miscegenation (the interbreeding of races) had been with him since his childhood" (Berger 1980, x). According to Michael Rogin, Twain concludes in this novel that biological heritage cannot be overcome by environment, for "blackening Tom made his 'one drop of Negro blood' (the Swedish title of the novel) into the sign of and explanation for his guilt" (Rogin 1990, 74). Rogin states that Tom "is no tragic mulatto, innocent victim," (74), rather, "Mark Twain conceived Tom as a coward and thief before he invented Tom's mulatto mother to account for Tom's character" (73–74). Consequently, Roxy's slave child—even with all the advantages of being raised as a white child—is deficient on many levels. Twain seems to be saying once again that whites are biologically superior human beings with greater potential—if they use it—for decency and honesty. While *The Descent of Man* corroborates this point of view, Darwin also states that environment is a major component in the modification of species. Twain seemed to be ambivalent about these influences on human development, but the naturalists embraced Darwin's message.

Without joining the debate about various definitions of literary naturalism (Conder 1984, 2), we want to focus here on the close correlation between Darwin's theory in *The Descent of Man* and American naturalism which *Merriam Webster's Encyclopedia of Literature* in 1995 defines as "a theory in literature emphasizing the role of heredity and environment upon human life and character development. . . . Naturalism differed from realism in its assumption of scientific determinism, which led naturalistic authors to emphasize the accidental, physiological nature of their characters rather than their moral or rational qualities. Individual characters were seen as helpless products of heredity and environment, motivated by strong instinctual drives from within, and harassed by social and economic pressures from without" (800). Clearly the Darwinian influences on naturalistic literary production are deep and have serious aesthetic consequences: the shared social Darwinist and eugenic belief in the inherent inferiority—or lack of development—of the "lower" orders of the human race; the notion that since humanity descended from our simian ancestors, then previous concepts of human origins and the meaning of life could well be fictions; and authorial acceptance of life based on chance, questionable causes and purposes, and scientific probability without divine teleology. Characters are drawn realistically and determined by contingency, inherited qualities, and adaptation to environment and circumstance. All behaviors must be accountable through rational analysis of the sources, or origins of the characters, as proliferating details about families, biology, and environment explain the motives and consequent actions of literary protagonists. Thus, we can see how American naturalism incorporated a wide range of influences like Marx and Freud, European writers like Zola, theories of social response and organizations, and Darwinian theories of animal origins, of adaptation to environmental forces, of inheritable traits in human descent with modification, and of the struggle of the fittest for survival. As Kenneth Rexroth wrote, "Naturalism and social Darwinism—these can be just ideological notions, or they can be reflections of a certain kind of human reality" (1964, 343). Both the "reality" and these ideologies are represented in much of American literature in the late nineteenth and twentieth centuries.

We see the theoretical issues and the reality played out in Stephen Crane's *Maggie, a Girl of the Streets* in 1893. The opening scene of the novel situates the characters in the jungle of Irish poverty in which the animal instinct for survival prevails. In the battle between Rum Alley and Devil's Row, between Jimmy and his foe, the children have not progressed beyond the Stone Age. They express themselves in "howls" like animals, "hurling stones and swearing in barbaric trebles" (Crane 1979, 3). The immigrants in this section of New York City are likened to animals—or less—as they exist "amid squat, ignorant stables" (3) and "a worm of yellow convicts . . . crawled slowly along the river's bank" (3–4). We hear "songs of triumphant savagery" (4) as the "small warriors" (5) remain locked into the age of our savage ancestors. Clearly, the rhetoric tells us what Crane thinks of the Irish Catholics and other immigrants populating the cities at that time as he judges and links them to the early progenitors of the putatively civilized human race.

In this novel, the individual's biology and cultural environment—including family, education, race, gender, economic and social class, religion, historical time, geographical place, and so on—condition Maggie to respond to psychological needs and experiences with results that cannot be altered by Maggie alone. Her family life, in which "animal qualities . . . predominate" (Pizer 1965, 188), her violent culture, her gender, and poverty reveal her presumed biological unfitness to survive. Only her derelict and brutal mother and brother, as well as Pete—Crane's picture of the worst specimens of the human species—survive to pass on their genetic heritage in the future. As a counter to Horatio Alger and other success novels and manuals in the nineteenth century, Crane's novel presents a dismal prediction of evolutionary "progress"—the social Darwinists' and eugenicists' worst nightmare.

Crane has gone far beyond Howells who, in *The Rise of Silas Lapham*, makes the distinction between ethical and material success and survival. Lapham ends by losing his wealth, but he maintains his integrity, his family's love and respect, and the small farm where he started his economic rise and where he now retreats from a world that overwhelms him. But in Crane's world there is no rural retreat or hope. The world of poverty and ignorance is a jungle where a gentle and loving child like Maggie is neither devious nor cruel enough to survive. No longer accepting Darwinian survival as defined by the number of one's offspring, writers now applied the concept of survival of the fittest to the individual's success in an industrialized, competitive, and brutal world. This version of success depended on defeating all opponents by physical prowess devoid of moral and ethical considerations. The biological struggle of evolutionism has become social and cultural, as the institutions of religion, family, school, and community are indifferent to the poor and helpless in their useless fight for survival. In this primitive landscape, human characteristics reflect qualities of the animal kingdom from which humanity has descended.

Frank Norris's *McTeague* (1899) is a primer on social Darwinism and determinism, as the language clearly indicates, for all characters are driven by habit and instinct and limited by the psychological, physical, and environmental forces that circumscribe their potential and their dreams. Trina and McTeague are attracted to each other by sexual drive and economic security, rather than rationality, and when that drive is satisfied and their sense of ownership no longer suffices, they have little in common and little love to share. They exist with the "little animal comforts" (Norris 1965, 220) until need and rage eventually destroy them. In their instinct for economic survival and in the competition between Marcus and McTeague for Trina, their animal natures surface and drive their behavior. All "human" qualities that evolved over millions of years, according to Darwin—qualities like compassion, loyalty, love of beauty, sense of community, and so on—are undermined by the animal natures that take over when poverty and hunger impel people to act. Norris presents a life in which the basic instinct to survive physically obviates ethics, morality, and the capacities for love, empathy, and nurture. It is a chilling vision of the competitive battle for survival in a world of limited resources and the struggle of each individual and species for what is available. It is this recognition of limited resources that made Darwin warn his readers about the reproduction of the poor and unfit who will take the food and land from those who are genetically the fittest for survival. Norris evidently acceded to Darwinian admonitions, for McTeague, Trina, and Marcus (as well as Maria and Zerkow) all die without issue. Darwin notes that there are "proofs that savages are independently able to raise themselves a few steps in the scale of civilisation" (1981, 1:181); however, Norris shows that the savage, primitive nature of human beings functions with a thin veneer of civilized behavior and that such "civilised" creatures revert to their primitive nature when they are tested by hunger, by the fear of death, and by the acquisitive desire for wealth to insure survival.

David Levinski, on the other hand, beats the competition by economic means. He proves that he is the fittest for survival both in the Old Country, where he endures terrible poverty and is orphaned as a result of anti-Semitic hatred, and later in the United States where he turns his superior intellect and Talmudic discipline into materialistic power and wealth. In *The Rise of David Levinski* (1917), Abraham Cahan deconstructs a capitalist system in which an immigrant's economic success, or the American Dream of riches and putative power through wealth, means complicity in violating all the values of religion, moral decency, and concern for the worth and feelings of others. Cahan suggests that achieving the American Dream in the early twentieth century requires both the willingness to enslave other human beings in the inhumane conditions of factory life and also a selfishness so profound that the fittest for success might well end up alone, unmarried, and without progeny to carry on his genetic line. In this novel on immigrant life, Cahan's socialist attack on the American economic system is clearly an attempt to interrogate and revise the social consequences of Darwinism.

Like so much of the literature of this time, an ideological thread of evolutionism runs through a plot that seems to be dealing with other issues. For example, David Levinski substitutes a secular philosophy of life for what had been a deep religious commitment before he came to the American shores. After his training to be a rabbi in the Old Country, he embraces new gods of education and wealth, and learns different lessons in New York: "The only thing I believed in was the cold, drab theory of the struggle for existence and the survival of the fittest. This could not satisfy a heart that was hungry for enthusiasm and affection . . ." (Cahan 1993, 380). At another point, he mentions the figures who most influenced him in his ascent to wealth in this country: "My interest in [Herbert Spencer] and in Darwin was of recent origin. . . . One day I found a long editorial in my newspaper . . . [that] derived its inspiration from the theory of the Struggle for Existence and the Survival of the Fittest. . . . 'Why, that's just what I have been saying all these days!' I exclaimed in my heart. 'The able fellows succeed, and the misfits fail. Then the misfits begrudge those who accomplish things.' I almost felt as though Darwin and Spencer had plagiarized a discovery of mine. . . . [With a vision of] chickens fighting for food . . . I had hit upon the whole Darwinian doctrine" (282). After reading "the *Origin of Species* and the *Descent of Man*, and then Spencer again" (282), Levinski knows his place in the universe: "Apart from the purely intellectual intoxication they gave me, they flattered my vanity as one of the 'fittest.' It was as though all the wonders of learning, acumen, ingenuity, and assiduity displayed in these works had been intended, among other purposes, to establish my title as one of the victors of Existence" (283). Like other members at the top of the class hierarchy, Levinski can look down on the lower orders of the human species: "A workingman, and every one else who was poor, was an object of contempt to me—a misfit, a weakling, a failure, one of the ruck" (283). Having arrived on these shores with a sense of his own intellectual superiority over the other immigrants because of his Talmudic education, he now has the scientific evidence to validate his own preeminence in a world of struggle and survival. Wealth is the evidence of his evolutionary superiority, and the price he pays for joining this "fittest" class has been to discard conscience, loyalty, family ties, and potential offspring (Perry 2000). According to Cahan, the capitalist definition of Darwinian "fitness" set the stage for the worst aspects of American materialism, commercialism, and narcissistic greed in the twentieth century.

The fusing of social class and economics to the principle of the survival of the fittest is a significant aspect of naturalism's ideology. This thesis is central to Theodore Dreiser's *An American Tragedy*, a novel often considered to be one of the prime examples of naturalism in American fiction. Clyde rejects family and traditional moral and religious values in his struggle for personal satisfaction and progress towards a higher class. To rise beyond the poverty and marginalized status of one's family is part of the American Dream, which motivates Clyde Griffiths in this novel. With proliferating details to account for the causes (or origins) of Clyde's thoughts, motives, and behavior, Dreiser constructs a protagonist whose sense of competition and survival leads him to murder a young woman in order to climb higher on the economic and social ladder. But within the Darwinian ideology, there is more to this situation. Clyde is conditioned by his heritage and environment to need communal attachment and power when he seduces Roberta; however, he eventually responds to Roberta's class "inferiority" by rejecting her despite their initial attraction and her pregnancy. At the same time, he is driven by an overpowering sexual instinct to mate with the "fittest" specimen when he sees Sondra whose genetic superiority (as her beauty, social class, and family wealth illustrate) makes her the more desirable mate for Clyde's biological survival and proof of Darwinian fitness. Dreiser further reveals that Clyde's weaknesses in character and family background, presumably the parental defects that led to their poverty, result inevitably in the choices that destroy him, despite his physical appearance and opportunities. By operating on the level of instinct and desire, he is the victim of his own primitive drives as well as his society's corrupt and misguided values.

The ending of the novel with Clyde's mother trying to "adapt" and thereby modify her own behavior in order to improve the next generation and the potential of the race might suggest that the future for her grandson will not mirror her son's life and death. However, a compelling argument can also be made that nothing else in the novel reveals such optimism because Darwinian progress is a long and slow process. Moreover, according to Lee Clark Mitchell, even the "repetitive patterns" in Dreiser's text refute any possibility of "progressive behavior" (1985, 40). Mitchell goes on, "the narrative patterns of *An American Tragedy* reveal how fully the motions of the self are wrenched into shape by an indifferent, and powerfully determinist, logic" (40). However we choose to read the ending, then, we can say only that this novel is exemplary in revealing the precise details that explain cause and effect in a mechanistic and deterministic world, the struggles and (un)fitness to survive, the competition and inability to adapt successfully, and the personal devastation that Charles Darwin with his privileged class, inherited money, and notions of organic survival and fitness could hardly have imagined.

Race and the Concept of Progress

Caucasian superiority and Negro inferiority were advanced by early American thinkers like Thomas Jefferson in 1786 and by subsequent apologists for slavery (Berzon 1978, 22). Later in *The Descent of Man*, Darwin furthered that agenda by documenting the close relation of lower animals to mammalian development and traced "the advancement of man from a former semi-human condition to his present state as a barbarian" to the condition of (white) man in "civilised nations" (1981, 1:167). With his premises and conclusions as a scientific foundation, racial issues and conflicts were directly related to social Darwinism and eugenic ideologies. On the one hand, Darwin placed the Anglo-Saxon peoples in the "higher" orders and affirmed by "scientific" proof the European (and American) white man's intellectual, moral, and physical superiority. He thereby justified white man's imperialistic goals, educational worthiness, and social and economic higher class. On the other hand, Darwin and his colleagues and followers relegated the world's non-Anglo peoples to the "lower" races associated with inferiority, barbarism, and a closer affinity to the primate progenitors of the human race. By 1900, a book entitled *The Negro a Beast* had "supernatural endorsement" for its racist cause in being published by a religious press (Berzon 1978, 28). Thus, there was a long history of racist rationalizations for American white superiority against which black writers have argued from the end of the nineteenth and throughout the twentieth centuries.

For African American writers and leaders, the issues regarding race and progress in this country required not only redefinitions of the construct of "race," but also social and political changes: equality between black and white citizens in terms of "political power," "civil rights," and "higher education for Negro youth" (Du Bois 1999, 40); refutation of assumptions of white superiority; the repudiation of conclusions about blacks as subhuman brutes versus civilized whites; exploration of passing, miscegenation and/or education as a path to progress; and so on. These were not playful fantasies or imagined subjects for fiction, but rather the very foundation of existence, rights, and freedom for the black population in the United States after the Civil War legally freed—but politically imprisoned—the former slaves. Often denied access to the American Dream simply on the basis of skin pigmentation, most people of color continued to be less educated, mired in poverty and hopelessness, cheated out of their land, and denied economic and political power to change their inferior position in society. Thus, literary realism and naturalism provided the means within popular culture by which African American authors could represent their lives and express the hope for human progress denied them by Darwinian and eugenic theories and by a society which embraced those principles.

Black intellectuals and leaders proposed a wide range of social changes, from accommodation to separatist power to miscegenation. Booker T. Washington, for example, proposed accommodation to white power by educating blacks to "dignify and glorify common labour, and put brains and skill into the common occupations of life" (Du Bois 1999, 168). Rejecting Washington's leadership in keeping blacks in a subservient position in the existing inequitable system, W. E. B. Du Bois called for "the rise of the Negro people, taking for granted that their best development means the best development for the world" (1999, xv). As the "first person of African descent to receive a Ph.D. from Harvard" in 1895 (xvi), Du Bois was "More than any other figure at the turn of the century . . . the public written 'voice' of the Negro American intellectual. . . ." (xvii) and remained so for decades.

Gates and Oliver document Du Bois's familiarity with the theory of evolution through his education at Harvard with professors like William James, George Santayana, and Barrett Wendell, "a social Darwinist" (1999, xix). Du Bois's writing is a response to the white intellectuals' complicity in the Darwinian and eugenic conclusions about the "lower races," which of course included African Americans, and their warning against "the weak members of civilised societies propagat[ing] their kind" (Darwin 1981, 1:168). Du Bois staked his claim for Negro humanity in *The Souls of Black Folk* by demanding "their status as humans *qua* humans in a world that would deny them this status" (Gates and Oliver 1999, xvii). In 1901 in "Of the Sons of Master and Man," he comments on white Western justification for imperial expansion and colonialism in the name of civilization and progress. Du Bois further situates his argument in the context of Darwinian principles: "It is, then, the strife of all honorable men of the twentieth century to see that in the future competition of the races the survival of the fittest shall mean the triumph of the good, the beautiful, and the true" (1999, 105–6).

No longer accepting survival in terms of biological reproduction alone, Du Bois looks to the qualities of morality, aesthetics, and spirituality that, according to Darwin's views, will enhance human life.

In "The Conservation of Races," a speech presented in March, 1897, Du Bois says that "out of one blood God created all nations" and that Negroes are "members of a vast historic race that from the very dawn of creation has slept, but half awakening in the dark forests of its African fatherland" (1999, 181). Yet, despite echoing the Darwinian concept that the source of life is shared by all races originally out of Africa, Du Bois preaches separatism and looks to "race unity," "race solidarity," and pride in the gifts and achievements of his people. He rejects "amalgamation," which would gradually eliminate Negro characteristics, and instead demands a new social, economic, and political environment to effect the changes that will raise up his race in the United States.

Like Alexander Crummell and other noted black leaders of the time, Du Bois reveals a style of thinking that crept into the psyche of intellectuals, regardless of their race—that is, a cognitive and rhetorical pattern that framed history and progress in terms of evolution. The difference, of course, is that Du Bois challenges the assumptions of white supremacy by first affirming both biological and social equality for his people; later, in his 1928 novel, *Dark Princess,* he envisions a new World Order in which the "darker world" is the "real" world of the future (1995, 246). Here Matthew learns about the realities of the world ("Dog eat dog is all I see," he says at one point [136]) and finally unites with Kautilya, the beautiful princess from India, whose vision will unite the world's people of color in the oneness of their single origin (227).

The novel documents the characters' sense of class and outrage over white racism, though people of color living elsewhere in the world also feel contempt for the "inferior" Negroes from America. Kautilya recounts to Matthew her experiences in Berlin where "they thought of American Negroes only as slaves and half-men" (247), and in England when an English boy rudely proposed marriage to her (231). Later her pain, rage, and contempt again surface when she recalls her betrothal to an English nobleman. She assumed that the impending marriage meant that she was considered equal to the English. Then she overheard her fiancé's conversation with the titled English woman he really loved:

> "Malcolm, I can't bear the thought of your mating with a nigger."
> "Hell! I'm mating with a throne and a fortune. The darky's a mere makeweight." (239)

By presenting the white, wealthy, and aristocratic family of "highest rank" (235) in the form of Malcolm—an unprincipled man whose motives for marrying this remarkable woman were to gain power over her state as "a pledge for future and wider empire" (239)—Du Bois comments on the corruption at the foundation of white racist hierarchy and imperialism. At the same time, by measuring this "poor and crippled younger son" (239) against the beautiful, intelligent, wealthy, and powerful princess, Du Bois deconstructs the baseless notions of white superiority that Darwin and Galton, both British men of privilege, promulgated. It is noteworthy, however, that Du Bois retains this class system at the end of his novel.

Dark Princess ends with an interesting message about progress. The articulated vision of a united, classless world of equality in color and gender becomes, in the novelist's hands, a patriarchal monarchy. As Kautilya says, "Bwodpur needs not a princess, but a King" (308). This novel subverts elements of its own message because there seems to be a conflict between the ideal and the reality, between what Du Bois believed and what he was taught to feel. In *Souls of Black Folk,* for example, the rhetoric of anti-Semitism—which he later denied (Gates and Oliver 1999, xxxix–xl)—remains a stain on his reputation and his ideas of equality. And in *Dark Princess,* despite his insistence on gender equality in the guise of Princess Kautilya, male dominance and patriarchy survive as the social and political model for the future. Du Bois simply substitutes blacks for whites in his paradigm of power. Like T. S. Eliot, Ralph Ellison, and so many authors and leaders in the twentieth century, Du Bois apparently could not free himself from the cultural baggage of his age. Thus, the novel ends with the hope of the future in a royal, male heir who disrupts the novelist's dreams of freedom and equality.

Black women writers were also influenced by Darwinian and eugenic ideologies and issues that relegated their people to inferior status in American society, but unlike Du Bois they did not always elevate the dark-skinned characters. A volume of *Short Fiction by Black Women, 1900–1920,* collected with introduction by Elizabeth Ammons, is a treasure of stories introducing readers to fiction published in the *Colored American Magazine* between 1900 to 1909 and *Crisis* between 1910 to 1920 (1991, 3). And while authors like Pauline Hopkins and Jessie Fauset incorporated political ideology into their fiction, women authors often confronted the questions of racial equality through plots of inadvertent or purposeful "passing" in which beautiful light-skinned women or men of "mixed blood" are superior to both blacks and whites. The narrative crisis occurs when the whites learn of the "tainted blood," and the responses of the characters are the measure of their worth in these stories.

Amalgamation was frequently promulgated as a method for achieving progress for the black races, and often in these stories, characters with "mixed blood" are shown to be superior physically and in terms of character. For example, the Darwinian, eugenic assumption of white man's physical and intellectual superiority is subverted in Annie McCary's "Breaking the Color-Line" in 1915. In this story Thacker, the fastest half-miler on the Starvard (Harvard?) track team, is rejected by his white teammates when they learn that he has "Negro blood in [his] veins" (Ammons 1991, 503): "he's a nigger, and no southern gentleman would compete against or run with a nigger"(502). Hearing these remarks, Thacker quits the relay team. But later he wins the race (and the track meet) for Starvard against Gale and thereby exhibits his superior physique, talent, and character, which refute Darwinian and eugenic laws of white supremacy.

Pauline Hopkins, author, literary manager of the *Colored American Magazine*, and one of the leading intellectuals of the time, was a prolific writer of short stories, essays, and four novels, which also depicted the trials and pain of life for African Americans. Much of her work directly addressed race in the context of social Darwinist and eugenic ideologies, as John Nickel's essay in this volume discusses. Nickel points out that the story of "Talma Gordon" expresses a common thread in much of Hopkins's work: that African Americans will benefit socially and politically from "amalgamation"—that is, intermarriage that will eugenically neutralize color distinctions by making blacks lighter. The language in this story is revealing because the person with mixed heritage must "possess decent moral development and physical perfection" in order to be worthy of the "pure Anglo-Saxon stream" (Hopkins 1991, 51). The concept expressed by the white doctor in terms of progeny is decidedly eugenic and suggests his (and perhaps Hopkins's) complicity in accepting the notion that "white" is somehow better. The interesting point is that in such intermarriage, the black female takes on the Anglo-Saxon color and characteristics as a sign that her heritage has been modified and improved. The Hegelian synthesis of two races results presumably in a Darwinian "higher" order when the progeny lose the signs of African descent. While Du Bois rejects that notion of a loftier status for Negroes by becoming white and pays tribute to the qualities of his own race, as did Hopkins later, authors of that time often defined female beauty in Anglo-Saxon terms.

In much of the African American literature around the turn of the century, women authors accepted the nineteenth-century notions of female beauty and thus often represented the most intelligent, talented, beautiful, and generous heroines as light-skinned, with "good hair." An example of this model is Burgess-Ware's "Bernice, the Octoroon," published in 1903, in which Bernice has "golden curls," a "blonde complexion, dainty mouth, and deep blue eyes" (Ammons 1991, 250). This beautiful, kind, and loving young woman has Anglo-Saxon features, presumably a sign of her preeminence as a human being. After growing up as white in a white family, Bernice learns that she is "one of a despised race" (251). And like most of the stories of inadvertent passing around the turn of the last century, the tainted "ancestry" came from the mother (a notable exception is Kate Chopin's "Desiree's Baby"), for the eugenicists believed that the mother's heritage was primarily responsible for the defective traits. For that reason, men were warned to select a mate wisely for the future of the family and the race. Garrett, like other lovers in these circumstances, has to make a choice; however, unlike so many others, Garrett remains loyal and loving to Bernice when she decides to make public her "alien race" and "despised blood of the Negro" (258). In an interesting narrative twist that secures the couple's future harmony, Garrett later learns that he too is a product of "tainted blood."

Burgess-Ware creates a narrator with strong eugenic attitudes, reflected particularly in Bernice's observations of the forty black children she has come to Maryland to teach:

> It never entered her pure mind that many of these knew no father; it did not dawn upon her that a race despising hers was at every opportunity flooding the country with children, born to be despised and persecuted. Some of these little ones were ragged and hungry, whose fathers lived in luxury, while *their mothers were ignorant women who knew nothing of the development of their intellectual being, but allowed their animal natures to predominate, and brought forth children regardless of the laws of God or man. Immorality to these*

> *people had no meaning*, and thus these poor little children's opportunities to become noble men and women, were very limited. (Ammons 1991, 261; emphasis added)

In this passage Bernice, through the narrator, blames the black women for the births that might well have been the result of rape and enforced servitude by the white men who "lived in luxury" and despised the women and their offspring for their race. The Negro women in poverty, in contrast to the "pure" golden Bernice who was brought up in wealth and privilege, are further castigated by being depicted with "animal natures" and procreating without the restraint of moral standards expected through evolution. Clearly, the narrator expresses a patronizing contempt for the poor and uneducated black women. Thus, like so many of the white authors, some of the black authors conflate race and class and thereby foster the hierarchical and racist doctrines of the evolutionists and eugenicists.

Nella Larson confronts these issues once again in the 1920s in *Passing* and *Quicksand*. With the recovery and publication of Nella Larson's two novellas, the conflict and ambivalence of a mulatto author during the Harlem Renaissance are opened up to us. In *Passing* (1929), for example, both racial and sexual hiding are played out (McDowell 1986, xiii–xiv). The problems involved in Clare's marriage to a white bigot, her alienation and ethical compromises, and the jealousy and resentment of the "black" women because of her color (in addition to the dangers of black female sexuality represented by Irene's jealousy and hostility) all conspire to destroy the beautiful Clare. Her name, a variation of the French word "claire," seems to be used ironically because her passing is not "transparent" or "clear" to the white community. However, Clare is "light-coloured" and "clear" in recognizing her feelings and her reasons for being who she is, while Irene is locked in a prison of repression and denial. Yet, Irene survives while Clare is punished by death apparently for turning against "her people" and living white. It is a complex novel in which the conscious and unconscious levels of each main character operate within a social context of "racial uplift" (McDowell 1986, xvi) that challenges everything but the "[i]nstinct of the race to survive and expand" (Larson 1986, 186).

However ambivalent the ending of *Passing* may be, *Quicksand* (1928) is even more distressing in Helga's complicity with culturally accepted attitudes about hierarchy and the value people place on color, about the alienation suffered by the person of mixed race who belongs nowhere, about the issues of black female sexuality, and about eugenic beliefs. With an interesting variation on the Darwinian and eugenic fears that the "fittest" white people will not reproduce enough to rule the world in the future, James expresses the same belief that he and Helga are the hope of *their* race: "'Don't you see that if we—I mean people like us—don't have children, the others will still have. That's one of the things that's the matter with us. The race is sterile at the top. Few, very few Negroes of the better class have children, and each generation has to wrestle again with the obstacles of the preceding ones, lack of money, education, and background. I feel very strongly about this. We're the ones who must have the children if the race is to get anywhere'" (Larson 1986, 103). After she breaks her engagement to James, Helga spends the next few years resisting this Darwinian, eugenic imperative until she comes back to the South to be trapped in the very condition of poverty and reproduction about which James warned her.

Larsen places this beautiful, intelligent woman in a linguistic, Darwinian context that reveals her conflicts about being black in a world that considers color to be related to our simian ancestors. In her hatred of all white people (48), particularly the "savage unkindness" of her white mother's new family (23), she chooses to live with the black upper class. Yet, she still feels ambivalent and alienated: "But she *aped* their clothes, their manners, and their gracious ways of living. While proclaiming loudly the undiluted good of all things Negro, she yet disliked the songs, the dances, and the softly blurred speech of the race. Toward these things she showed only a disdainful contempt. . . ." (48; emphasis added). The Harlem jazz club is a particular threat, for the "wild, murky orchestra" with its "savage strains of music" stirs in her the "essence of life": "And when suddenly the music died, she dragged herself back to the present with a conscious effort; and a shameful certainty that not only had she been in the jungle, but that she had enjoyed it, began to taunt her. She hardened her determination to get away. She wasn't, she told herself, a jungle creature" (59). All about her in this club were varied shades of skin and eye colors, different textures of hair, "in a fantastic motley of ugliness and beauty, semi-barbaric, sophisticated, exotic. . . ." (59–60). Influenced by both European and American societies that "pandered to the stereotype of the primitive exotic" (McDowell 1986, 16) and of black woman's animal sexuality, Helga fears her own "savage" nature. She then rejects black culture until the moment in the African American church when the sensuality of the music and the seductive rhythms of the minister's sermon break the barriers of desperation and loneliness. Rationality and resistance are replaced with desire and the need to lose herself in someone else's strength and faith. Ironically, it is this moment of freedom that traps her forever.

In the references to "descent" in this novel, we watch Helga become mired in the quicksand of Darwinian, eugenic determinism based on birth, race, class, and ultimately gender. Even after she marries the black minister and returns to the South, she feels contempt for black culture and particular disgust because of the poverty and lack of education of those around her. Fulfilling the prophecy of James, she has one baby after the other, is unable to cope with the physicality, dirt, and hard work required of a poor country minister's wife, and descends both in class and in lost opportunity. This creates a psychological morass that immobilizes her. Having loved beautiful things and having needed absolute control over her body and her life, she now hates the ugliness of poverty, her husband, the people around her who can laugh at life, and even her children, though she knows she "couldn't desert them" (135). Bound by the expectations for women and by the bitter choices linked to sexuality and birth for black women, Helga is caught in an inescapable circle that was prescribed by her beginnings: "And hardly had she left her bed and become able to walk again without pain, hardly had the children returned from the homes of the neighbors, when she began to have her fifth child" (135). There is no promise of progress here. And we see little hope of progress in African American literature even in the decades following the Second World War.

Modernism and the Paradigm of Descent

Like the categories of romanticism, realism, naturalism, and other literary movements, modernism is difficult to define with generalizations. Moreover, there is no exact beginning or end because literary periods overlap and are modified by cultural changes. However, some common influences and characteristics may help us to understand the philosophical and aesthetic framework that became the world view of many authors in the first part of the twentieth century, authors as different as T. S. Eliot, Eugene O'Neill, and Djuna Barnes. In their rebellion against realism and scientism as truths that replaced the pre-Darwinian absolutes, modernist authors were often concerned less with realistic details and facts, with exploring causes and effects, or with acting out the survival of the fittest principle in their work. Instead, they focused on finding the metaphors, abstractions, and unconventional language and form (Bradbury and McFarlane 1981, 24) that could describe the human condition as they saw it. Clearly, the theory of evolution resulted in the disintegration of old ideas (*Life* 1958), and Bradbury and McFarlane describe Modernism in terms that echo this profound impact on Western civilization: "those overwhelming dislocations, those cataclysmic upheavals of culture, those fundamental convulsions of the creative human spirit that seem to topple even the most solid and substantial of our beliefs and assumptions, leave great areas of the past in ruins (noble ruins, we tell ourselves for reassurance), question an entire civilization or culture, and stimulate frenzied rebuilding" (1981, 19). Modernism, they write, deals with "experimentalism . . . [that] suggests bleakness, darkness, alienation, disintegration. . . . it often involves an unhappy view of history. . . . an awareness of contingency as a disaster in the world of time" (26):

> Modernism is the art consequent on Heisenberg's "Uncertainty principle", of the destruction of civilization and reason in the First World War, of the world changed and reinterpreted by Marx, Freud and Darwin, of capitalism and constant industrial acceleration, of existential exposure to meaninglessness or absurdity. It is the literature of technology. It is the art consequent on the dis-establishing of communal reality and conventional notions of causality, on the destruction of traditional notions of the wholeness of individual character, on the linguistic chaos that ensues when public notions of language have been discredited and when all realities have become subjective fictions. . . . [It is] the expressive form of human evolution in energetic release. (27)

This, briefly, is the epistemology of modernism, which scholars like Bradbury and McFarlane (1981) have traced in European intellectual production into the twentieth century. Yet, despite our knowledge of these major influences, despite Darwin's "special and outstandingly influential role" in the intellectual life of his age, and despite the acknowledgment of his theory as "one of the seminal ideas in the history of thought" (73), the extent and treatment of evolution and eugenics in American modernism still require further examination.

Unlike the naturalists, who applied scientific principles directly to their representations of human life in their work, many modernists used evolution as an organizing principle for a Western world view of descent and development, for the present was based on the past: on the sources and origins that explain the present in terms of individual and cultural development, on the stages of the Earth and its species, and on the cosmos and its relation to life on Earth. All things were related because everything evolved from a common source, and literature reflected the nature of such relations. T. S. Eliot, the Nobel poet and the literary voice for his age, may be one of the clearest examples of how the theory of evolution informed aesthetic philosophy in modernism and resulted in a poetics of unity designed to resist disintegration and to restore order in life.

Born in 1888 in St. Louis, Missouri, Eliot was clearly a product of his time and education. Thus, in his work, the theory of evolution in society became wedded to the Western classical tradition, or the Homeric/Odyssean tradition, in academia. Scholars have thoroughly documented these influences through his mother's interest in evolution, his education at Harvard where evolutionary theory was a common topic for debate and for the organization of course syllabi, through the writings of his friends and professional colleagues, through his knowledge of anthropology and eugenics (Crawford 1987; Leon 1988; Manganaro 1992), and through his extensive knowledge of the classics generally and of Homer specifically (Smith 1974, Cuddy 2000). Critics have generally interpreted Eliot's poetic style of "borrowing" from authors across the ages as an exclusionary and elitist tactic to show his own intellectual superiority as well as the disjunction and fragmentation of the modern world. However, beneath the social chaos that he describes in *The Waste Land* and other poems, Eliot juxtaposes his poetic forms and allusional strategies with the theories of evolution, anthropology, and eugenics—particularly Charles Darwin's principle of "descent with modification"—to create his grand design for the unity and evolution of the human race and of existence as he had come to know it.

To achieve this aesthetic and thematic design of cultural evolution, Eliot wove throughout his poems and plays literary, philosophical, mythical, and religious allusions from Western history and Eastern writings. Particularly he used Homer, long considered the "father" or origin of Western literature and ideas, as a primary reference point as he traced numerous authors' treatment of Odysseus and of Homeric images and themes to reinforce his sense of cultural history. He then juxtaposed those ubiquitous Odyssean allusions with references to Darwinian, cultural, cosmic, and personal evolution. With these methods, he attempted to map the history of human ideas and behavior, the heroic tradition, family and gender relationships, cultural values, and his own development.

For Eliot, Darwinian "descent" became a metaphoric umbrella, which covered all ideas and human expression through time. It provided the model for Eliot's physiological "descent" of humanity from the "protozoic slime," to "Apeneck Sweeney," to the poet's own life (Eliot 1952, 24, 35); it also offered a trope for Western tradition's spatial "descent" into the Underworld in both literature and religion; and it suggested Eliot's metaphoric "descent" into his own Hell. It is the "descent" of humanity that Eliot's poetry and dramas attempt to summarize. Thus, the union of evolution, history, and myth became Eliot's definition of "Western Tradition" from the pre-Homeric world to the present.

Eliot critiqued the evolutionists' belief in progress and concluded that there had been little improvement in the human race since before the time of Homer, except in material and scientific advances. The warnings of Darwin and the eugenicists had taken shape in the societies of Europe and the United States, so that the "descent" of man had become a literal regression in the human race. Like so many of his figures, "Sweeney Erect," Prufrock, and Gerontion illustrate the decline of the heroic tradition and provide evidence that man's development of conscience, moral principles, community, and literacy—qualities that Darwin pointed to as signs of the highest level of progress—have only served to torment the most intelligent, sensitive, and creative members of our species and to deny these men the peace and happiness that "man" has sought through time.

Like other modernists, Eliot's own limitations in accepting Darwin's hierarchy and relegating the workers, the poor, women, and other racial and ethnic groups into a lower evolutionary realm, which he observed from the pinnacle reserved for very few men, resulted in a psychological trap that undermined the poet's mission of unity and sense of progress. Yet, in his singular version of evolution as a principle of life and art, in his struggle with spiritual belief and with language—which was part of the linguistic branch of evolutionary science—and with his critique of progress, Eliot spoke for his age and for modernist authors as diverse as Conrad, Joyce, Virginia Woolf, HD, Faulkner, and O'Neill.

After publication of *The Descent of Man* in 1871 when Darwin stated clearly that man gained his parentage from the "Old

World monkeys" (1981, 1:213), the correlation of man and ape became part of the American psyche. In 1875, for example, in an article entitled "Our Ape Relations" on the first page of *The Home Journal*, a St. Louis newspaper, a reporter chided readers who resisted Darwin's theory by pointing out "the imitative instinct" that is everywhere evident in human as well as simian social groups. The article was intended as a moral lesson directed at smoking and drinking alcohol, habits that reflect the "spirit of imitation" inherited from our "brutish prototype." As literature and popular culture came to accept the link between humanity and our animal progenitors, then, this identification had many different literal and metaphoric applications. One of the most explicit appropriations of Darwin's observation of the ape/human relationship is in the 1921 play, *The Hairy Ape,* by Eugene O'Neill.

In that drama, O'Neill uses the notion of our evolutionary origin from the primates to create a bleak vision of life for humanity in the modern world. It is a modernist landscape of alienation, class hatred, and the consequences of biological inheritance as the protagonist and his mates working in the stoke hole of an ocean liner are depicted as animals. Their primitive lives, behavior, and appearance as they work like the animals that they are alleged to be link them directly to their Darwinian forebears: they wait in the "crouching, inhuman attitudes of chained gorillas" (O'Neill 1954, 55); Yank has a "gorilla face" as he responds with a "snarling, murderous, growl" and with "one hand, pounding on his chest, gorilla-like" (58); Paddy complains that in bringing Mildred into the hold they are "exhibitin' us 's if we was bleedin' monkeys in a menagerie" because "she wants to see the bloody animals below decks"; and repeatedly we hear the men in "a chorus of hard, barking laughter" as they swallow insults "like dogs" (61). Paddy gives Yank a new and horrifying self-image and identity when he says, "In this cage is a queerer kind of baboon than ever you'd find in darkest Africy[9]. . . . Sure, 'twas as if she'd seen a great hairy ape escaped from the zoo" (62). This drama focuses on Yank, the Hairy Ape, when he suddenly sees himself through the eyes of Mildred, the spoiled remnant of a dissipated wealthy class.

In this drama, O'Neill shows the consequences of the "descent" of the human race for both the upper and under classes. We see that the wealthy have lost their hearts and conscience and their sense of life's meaning in their perpetual drive for material goods, and the workers have lost their humanity in the service of the rich and powerful. These workers are treated with contempt and remain caged by that very industrialism of steel and bars that keeps them living like animals. Yank represents this defeat and denigration. Rejected for his violent rage against everyone and everything because of who he is, Yank cannot find a place to "belong" among people and goes to the "monkey house at the Zoo" (84). Here the gorilla sits in the posture of Rodin's "The Thinker," a position that Yank often ironically assumed, and Yank says to the curious animal, "Ain't we both members of de same club—de Hairy Apes?" (85). Yet Yank, both gorilla and man but neither one completely, belongs nowhere and to nothing until he dies. Playing on eugenics and the prevailing theories on race, O'Neill depicts the regression of the white race as the white laborer is likened to his simian ancestor. He thereby lives out his life in frustrated yearning for what he will never have or be and in pain for what he is.

Again in *Nightwood* (1936), Djuna Barnes creates a human race caught between the requirements for civilized behavior and the instincts that drive their animal natures. Robin is a protagonist who is more animal than human, who is incapable of bonding on any but the most sexually primitive levels, who prowls the bars and streets at night for brief sexual encounters, and who ultimately descends in behavior and utterance to the literal level of the dog. The past, our origin and source, is within her, for Robin represents the history of life on earth: her flesh is "the texture of plant life," and "her body exhaled . . . the quality of that earth-flesh, fungi. . . . and . . . oil of amber, which is an inner malady of the sea" (Barnes 1961, 34). With a beauty that seems human and a laugh from "some lost subterranean humor" (47), Robin "lives in two worlds" (35). The narrator introduces this figure with a metaphoric setting that forecasts her transhistoric natures: "she seemed to lie in a jungle trapped in a drawing room" (35), her eyes having "the long unqualified range in the iris of wild beasts who have not tamed the focus down to meet the human eye" (37). While Nora is both "savage and refined" (50) and Felix is both man and animal (32, 123), Robin is also animal and desirable woman as she survives without human conscience but with the instincts that destroy others. Felix has made a poor eugenic choice in mating with her in the expectation that she will strengthen his "Jewish race," as the weaknesses of his son apparently prove. Except for maternal affection that Robin lacks but that Darwin believed to be universal in human and animal for survival (Darwin 1981, 1:40), Robin illustrates other elements of Darwin's theory, particularly about how "intimately related" (1:130) the lower animals (like the dog) and human beings really are (1:40, 66).

Like Darwin's construct of "descent with modification" in which chance and environmental forces unite with genetics to alter species over time, the world of this modernist novel is in constant flux. Nothing is stable or certain: nothing is as it seems in this "sad and . . . corrupt age" (161). People are not what they present themselves to be, even as time, myth, titles, traditional values, habits, and sexuality are turned upside down in this novel of night and shadow, the ancient past and the present. It is a surreal world framed by a transformed allegory of evolution that focuses on the "descent" of humanity rather than the upward progress that cosmic evolutionists promulgated (Fiske 1893; Boodin 1925). Darwin ended *The Descent of Man* with pride in white, "civilised" man's "noble qualities" and "god-like intellect which has penetrated into the movements and constitution of the solar system—[Yet] with all these exalted powers—Man still bears in his bodily frame the indelible stamp of his lowly origin" (2:405). It is the "stamp of his lowly origin" which the modernist authors explore—the primitive underworld of the psyche, instincts, desire, violence, and despair as they observe their world and conclude that humanity has made little progress in its attempt to alter the nature of what we have become.

Transforming conventional patterns and metaphors of myth, history, time, tradition, and literary genres, much of modernism appropriates then subverts the evolutionary model by representing the human race lost in a wasteland. It is not realism in terms of everyday details and events, but rather a sense of the reality of life, for "formed in man's image is a figure of doom" (Barnes 1961, 41). By the 1930s, then, Darwin's insistent optimism about the possibilities for the human race and the eugenic demands that society could create perfect human beings had lost all currency for authors living in a world of economic depression and human deprivation and suffering. As the modernists, socialists, and other people of sensitivity and conscience looked at their societies in Europe and America, the horrors of the human condition could hardly be called progress.

Conclusion

As we look back on the period roughly between 1880 and 1940, we realize that the concept of human progress remained wedded to Darwinian and eugenic prejudices. And regardless of the literary form in which it is presented, the Darwinian way of seeing the world and human life had taken root in both European and United States literature and culture. Even intellectuals who, from our contemporary perspective, should have rejected the premises of white, Christian, heterosexual, male superiority could not free themselves from such biased conclusions. In the Western countries, science had empowered upper-class, educated, white men to enjoy the only thing they could believe with absolute certainty: their own preeminence in a world of constant change. And for the most part, women and blacks, hearing this message repeated in multiple ways at every moment of their lives, consciously or unconsciously appropriated that construct and made it part of their own psyche. Thus, even many women and African American authors, while attempting to resist Darwinian and eugenic principles, were powerless to reconfigure life in images other than those reflected in a distorted cultural mirror. Little wonder that the literature of the later nineteenth and twentieth centuries reflects the powerful messages of Darwin, Galton, Davenport, and their advocates and maps the disturbing biases in American culture during that period.

Yet, despite the pervasive and profound influence of evolution and eugenics on American society, history, and literature between *The Descent of Man* and World War II and despite the interest centering on these issues during the final decades of the last century, much is yet to be done on these subjects in literary criticism. This volume of essays on a wide range of authors provides compelling evidence that we must extend even further our reexamination of literature in this historical and social context. Therefore, in this volume we have included essays on both canonical and noncanonical, black and white, female and male authors,[10] on authors who were popular in their time and on those who have remained favorites, on different literary genres, different geographical settings and publication venues, and different theoretical positions—essays that provide a view of an American literary landscape shaped by social attitudes and ideologically based laws. Furthermore, we wanted to explore the vast range of influences of these theories on American society and literature and to ask readers to consider the role of literature in social structures, attitudes, and beliefs: whether, in fact, authors represented and replicated entrenched cultural biases or whether authors are ahead of their time and lead their societies in new ideational directions in the interest of progress. We believe that the essays in this volume go far in exploring many of these intellectual and scholarly issues, and we hope that this study generates the interest to open up new research and perspectives on these subjects.

❧

Note: For the convenience of readers, a list of references, or works cited, follows each essay in the volume. All together, these bibliographies offer a wide range of citations on the topics of evolution and eugenics, as well as sources on the particular subjects of each individual essay.

Notes

1. The last word in *The Origin of Species* is "evolved"; however, Herbert Spencer is credited with naming Darwin's theory "evolution."

2. According to James Sully in the ninth edition of the *Encyclopaedia Brittanica* (1878), the belief in some form of evolution is as old as ancient Greek philosophy. Moreover, Darwin's grandfather, Dr. Erasmus Darwin, suggested the general outline in *Zoonomia* for the theory of sexual selection that Alfred Wallace and Charles Darwin published in 1858 and that Charles later documented in detail (Irvine 1955). However, it was *The Origin of Species* that offered to the world, with enormous evidence and detail, the grand scheme of descent with modification that changed Western thinking.

3. We use the term "man" when discussing Darwin and other evolutionists because it is consistent with the gendered diction in their texts and in the time period that is the purview of this volume. Moreover, while concepts of race have changed quite dramatically since the time period that we are discussing, and Jews, for example, are no longer considered a race, we have tried to compromise on the language of the time in order to be consistent with the authors being discussed.

4. Despite the vilification of Darwin by the religious establishment and people who had never read what the scientist had actually written, Darwin did not deny the existence of God or a force that "breathed life" into the first primordial cell (Darwin 1985, 455); rather, he had lost all orthodox belief and concluded that "questions of ultimate causes and purposes were an insoluble mystery" (Burrow 1985, 24).

5. Donald Pizer notes, "Many people were more comfortable with the evolutionism of Herbert Spencer whose "Synthetic Philosophy was grounded upon evolutionary science, but it also acknowledged both the existence of an 'unknowable' beyond human perception and the evolution of man's psychic nature—or soul—as well as his body" (1972, 1).

6. Judith R. Berzon notes that "Social Darwinism itself was a perversion of Darwinian thought" (1978, 27). She quotes the words of George Fredrickson: " 'By appealing to a simplistic Darwinian or hereditarian formula, white Americans could make their crimes against humanity appear as contributions to the inevitable unfolding of biological destiny' " (27). Though Charles Darwin played

no part in the formulation of social Darwinism and in how his theories were applied by social scientists and eugenicists, the movement gained credence by the participation of his son, Major Leonard Darwin, as president of the British Society of Eugenics from 1911 to 1928 (Kevles 1986, 60).

7. What John Taliaferro says about Edgar Rice Burroughs and his creation of the Tarzan legend has application to many of the authors writing in the stream of popular culture in the late nineteenth and early twentieth centuries:

> "Burroughs' life-long fascination with evolution [was] an interest he shared with millions of his contemporaries. . . . As the physical [archeological and anthropological] evidence accumulated, the links in the evolutionary chain suddenly seemed finite, almost countable. Now that modern man was able, in effect, to look his forefathers in the face, the gap [between Neanderthal man and modern man] didn't seem all that vast. Robert Louis Stevenson imagined that within every Jekyll was a potential Hyde. Jack London, an author whom Burroughs admired enormously, argued in a shelf of evocative novels that the 'call of the wild' was still echoing in the inner ear of all domesticated animals, including humans. Similarly, Tarzan possesses 'the best characteristics of the human family from which he descended and the best of those which mark the wild beasts.'" (1999, 92–93)

Later in the volume, Taliaferro discusses Burroughs's belief in eugenics, for "If creationism was the antithesis of Darwinism, then eugenics was Darwinism taken to extremes" (226ff.). Burroughs's assertions that "moral imbeciles breed moral imbeciles, criminals breed criminals," and so on (230), were articulated especially in his writings on the Hickman trial and in his novel, *Lost Empire* (Taliaferro 1999, 230–31). Like so many authors of his time, Burroughs selected from these complex theories only the details and concepts that he could use (and manipulate) to express his own views on life.

I wish to thank Prof. Sue Vaughn and Robert Vaughn for pointing out this source and for sharing their time and their library resources with me.

8. Scholars have verified Twain's interest in Darwinism and social Darwinism. See, for example, Stan Poole, 1985, "In Search of the Missing Link: Mark Twain and Darwinism" (201–15).

9. In *The Descent of Man*, Darwin stated, "It is therefore probable that Africa was formerly inhabited by extinct apes closely allied to the gorilla and chimpanzee; and as these two species are now man's nearest allies, it is somewhat more probable that our early progenitors lived on the African continent than elsewhere" (1981, 1:199).

10. While there are references to sexuality in relation to evolution and eugenics in this volume, much needs to be done on the subject, especially in light of the often unstated assumption that only heterosexuals, in terms of reproduction, are relevant to this discussion.

References

Achebe, Chinua. 1989. "An Image of Africa: Racism in Conrad's *Heart of Darkness*." In *Hopes and Impediments: Selected Essays*. New York: Doubleday.

Ahnebrink, Lars. 1961. *The Beginnings of Naturalism in American Fiction, 1891–1903*. New York: Russell and Russell.

Almog, Shmuel, ed. 1988. *Antisemitism Through the Ages*. Translated by Nathan H. Reisner. Oxford: Pergamon Press.

Ammons, Elizabeth, ed. 1991. *Short Fiction by Black Women, 1900–1920*. The Schomburg Library of Nineteenth-Century Black Women Writers. New York: Oxford University Press.

Barnes, Djuna. [1937] 1961. *Nightwood*. Reprint, with an introduction by T. S. Eliot. Second American edition. New York: New Directions.

Berger, Sidney E., ed. 1980. *Pudd'nhead Wilson* and *Those Extraordinary Twins* by Mark Twain. 1892. Reprint, 2 volumes in 1. New York: W. W. Norton.

Berzon, Judith R. 1978. *Neither White Nor Black: The Mulatto Character in American Fiction*. New York: New York University Press.

Blake, Lillie Devereux. [1874] 1996. *Fettered for Life, or Lord and Master*. Reprint, edited with an afterword by Grace Farrell. New York: The Feminist Press at The City University of New York.

Boodin, John Elof. [1925] 1970. *Cosmic Evolution: Outlines of Cosmic Idealism*. Reprint, New York: Kraus.

Bradbury, Malcolm and James McFarlane, eds. 1981. *Modernism, 1890–1930*. Pelican Guides to European Literature. Harmondsworth, England: Penguin Books.

Budd, Louis J. 1995."The American Background." In *The Cambridge Companion to Realism and Naturalism*, edited by Donald Pizer. Cambridge, Mass.: Cambridge University Press, 21–46.

Bureau of the Census. Census 2000. www.census.gov/main/www/cen2000.html (12 November 2000).

Burgess-Ware, M.[arie] Louise. [1903] 1991. "Bernice, The Octoroon." *Colored American Magazine* 6: 652–57. Reprint, in *Short Fiction by Black Women, 1900–1920*, edited with an introduction by Elizabeth Ammons. New York: Oxford University Press.

Burrow, J. W., ed. [1859] 1985. *The Origin of Species by Means of Natural Selection or The Preservation of Favoured Races in the Struggle for Life*. Reprint, London: Penguin Classics.

Cahan, Abraham. [1917] 1993. *The Rise of David Levinsky*. Reprint, with an introduction by Jules Chametzky. New York: Penguin Books.

Conder, John J. 1984. *Naturalism in American Fiction: The Classic Phase*. Lexington: University Press of Kentucky.

Crane, Stephen. [1893] 1979. *Maggie: A Girl of the Streets*. Reprint, edited by Thomas A. Gullason. A Norton Critical Edition. New York: W. W. Norton.

Crawford, Robert. 1987. *The Savage and the City in the Work of T. S. Eliot*. Oxford: At the Clarendon Press.

Cuddy, Lois A. 2000. *T. S. Eliot and the Poetics of Evolution: Sub/Versions of Classicism, Culture, and Progress*. Lewisburg, Pa.: Bucknell University Press and London: Associated University Presses.

Darwin, Charles. [1871] 1981. *The Descent of Man, and Selection in Relation to Sex*. Reprint, 2 volumes in 1, with an introduction by John Tyler Bonner and Robert M. May. Princeton, N.J.: Princeton University Press.

———. [1859] 1985. *The Origin of Species by Means of Natural Selection or The Preservation of Favoured Races in the Struggle for Life*. Reprint, with an introduction by J. W. Burrow. London: Penguin Classics.

———. "The writings of Charles Darwin on the web," compiled by John van Wyhe. http://pages.britishlibrary.net/charles.darwin/

Dreiser, Theodore. [1925] 1981. *An American Tragedy*. Reprint, with an afterword by Irving Howe. New York: New American Library.

Du Bois, W. E. B. [1928] 1995. *Dark Princess, A Romance*. Reprint, with an introduction by Claudia Tate. Jackson: University Press of Mississippi.

———. [1903] 1999. *The Souls of Black Folk*. Reprint, edited and with an introduction by Henry Louis Gates Jr. and Terri Hume Oliver. A Norton Critical Edition. New York: W. W. Norton.

Eliot, T. S. 1952. *The Complete Poems and Plays, 1909–1950*. New York: Harcourt, Brace & World.

Encyclopaedia Britannica: A Dictionary of Arts, Sciences, and General Literature. 1878. "Evolution," by James Sully and T. H. Huxley. 9th edition. New York: Charles Scribner's Sons 8: 744–72.

Farrell, Grace, ed. 1996. *Fettered for Life* by Lillie Devereux Blake. 1874. New York: The Feminist Press of The City University of New York.

Fiske, John. 1893. *Outlines of Cosmic Philosophy Based on the Doctrine of Evolution, with Criticisms on the Positive Philosophy*. Boston: Houghton Mifflin.

———. [1899] 1972. "The Doctrine of Evolution: Its Scope and Purport." In *A Century of Science and Other* Essays. Reprint, in *American Thought and Writing: The 1890's*, edited by Donald Pizer. Boston: Houghton Mifflin.

Gallagher, Nancy L. 1999. *Breeding Better Vermonters: The Eugenics Project in the Green Mountain State*. Hanover, N.H.: University Press of New England.

Gates, Henry Louis, Jr., and Terri Hume Oliver, eds. 1999. Introduction to *The Souls of Black Folk* by W. E. B. Du Bois. Reprint, A Norton Critical Edition. New York: W. W. Norton.

Holmes, Sarah. 2002. "Leftist Literature and the Ideology of Eugenics During the American Depression." Ph.D. diss., Department of English, University of Rhode Island.

Hopkins, Pauline E. [1900] 1991. "Talma Gordon." *Colored American Magazine* 1: 271–90. Reprint, in *Short Fiction by Black Women, 1900–1920*, edited with an introduction by Elizabeth Ammons. New York: Oxford University Press.

Howells, William Dean. [1885] 1980. *The Rise of Silas Lapham*. Reprint, New York: Penguin Books.

Huxley, Sir Julian. 1958. "Darwin Discovers Nature's Plan." Part 1 in the series "The Wonders of Life on Earth." *Life* 44, 30 June.

Inge, W. R. 1909. "Some Moral Aspects of Eugenics." *Eugenics Review* (April): 1–3.

Irvine, William. 1955. *Apes, Angels, & Victorians: The Story of Darwin, Huxley, and Evolution*. New York: McGraw-Hill.

James, William. [1895] 1972. "Is Life Worth Living?" In *International Journal of Ethics* 6: 1–24. Reprint, in *American Thought and Writing: The 1890's*, edited by Donald Pizer. Boston: Houghton Mifflin.

Kevles, Daniel J. 1986. *In the Name of Eugenics: Genetics and the Uses of Human Heredity*. Berkeley and Los Angeles: University of California Press.

Kidd, Benjamin. 1894. *Social Evolution*. New edition. New York: Macmillan.

Larsen, Nella. 1986. [1928] *Quicksand* and [1929] *Passing*. Reprint, 2 volumes in 1, with an introduction by Deborah E. McDowell. New Brunswick, N.J.: Rutgers University Press.

Leon, Juan. 1988. "'Meeting Mr. Eugenides': T. S. Eliot and Eugenic Anxiety." *Yeats Eliot Review* 9: 169–77.

Manganaro, Marc. 1992. *Myth, Rhetoric, and the Voice of Authority: A Critique of Frazer, Eliot, Frye, and Campbell.* New Haven, Conn.: Yale University Press.

McCary, Annie. [1915] 1991. "Breaking the Color-Line." *Crisis* 9: 193–95. Reprint, in *Short Fiction by Black Women, 1900–1920*, with an introduction by Elizabeth Ammons. New York: Oxford University Press.

McDowell, Deborah E., ed. 1986. Introduction to *Quicksand* [1928] and *Passing* [1929] by Nella Larson. Reprint, 2 volumes in 1. New Brunswick, N.J.: Rutgers University Press.

Merriam Webster's Encyclopedia of Literature. 1995. Springfield, Mass.: Merriam Webster.

Mitchell, Lee Clark. 1985. "And Then Rose for the First Time": Repetition and Doubling in *An American Tragedy.*" *Novel* (Fall): 39–56.

Norris, Frank. [1899] 1964. *McTeague, A Story of San Francisco.* Reprint, with an afterword by Kenneth Rexroth. New York: New American Library.

O'Neill, Eugene. [1921] 1954. *The Hairy Ape.* In *Nine Plays by Eugene O'Neill.* New York: The Modern Library.

Perry, Matthew S. 2000. "For the Love of Money: The Representation of Capitalism in American Popular Texts." Master's thesis, Department of English, University of Rhode Island.

Pizer, Donald. 1965. "Stephen Crane's *Maggie* and American Naturalism." *Criticism, A Quarterly for Literature and the Arts* 7: 168–75.

———, ed. 1972. Introduction to *American Thought and Writing: The 1890's.* Boston: Houghton Mifflin.

———, ed. 1998. *Documents of American Realism and Naturalism.* Carbondale and Edwardsville: Southern Illinois University Press.

Poole, Stan. 1985. "In Search of the Missing Link: Mark Twain and Darwinism." *Studies in American Fiction* 13.2: 201–15.

Rexroth, Kenneth. 1964. Afterword to *McTeague: A Story of San Francisco* by Frank Norris. New York: New American Library.

Riis, Jacob. [1890] 1971. *How the Other Half Lives: Studies Among the Tenements of New York.* Reprint, with a preface by Charles A. Madison. New York: Dover.

Rogin, Michael. 1990. "Francis Galton and Mark Twain: The Natal Autograph in *Pudd'nhead Wilson.*" In *Mark Twain's* Pudd'nhead Wilson: *Race, Conflict, and Culture*, edited by Susan Gillman and Forrest G. Robinson. Durham, N.C.: Duke University Press, 73–85.

Rosenberg, Charles. 1997. *No Other Gods: On Science and American Social Thought.* Baltimore, Md.: Johns Hopkins University Press.

Schweber, Silvan S. 1985. "The Wider British Context in Darwin's Theorizing." In *The Darwinian Heritage*, edited by David Kohn. Princeton: Princeton University Press.

Shapiro, Thomas M. 1985. *Population Control Politics: Women, Sterilization and Reproductive Choice.* Philadelphia: Temple University Press.

Shi, David. 1995. *Facing Facts: Realism in American Thought and Culture, 1850–1920*. New York: Oxford University Press.

Shipman, Pat. 1994. *The Evolution of Racism: Human Differences and the Use and Abuse of Science*. New York: Simon and Schuster.

Smith, Grover. 1974. *T. S. Eliot's Poetry and Plays: A Study in Sources and Meaning*. Second edition. Chicago: The University of Chicago Press.

Taliaferro, John. 1999. *Tarzan Forever: The Life of Edgar Rice Burroughs, Creator of Tarzan*. New York: Scribner.

The Home Journal. 1875. No. 41 (13 October): 1.

Trachtenberg, Alan. 1982. *The Incorporation of America: Culture and Society in the Gilded Age*. New York: Hill and Wang.

Twain, Mark [Samuel Langhorne Clemens]. [1884] 1962. *Adventures of Huckleberry Finn (Tom Sawyer's comrade)*. Reprint, with an introduction by Hamlin Hill. Chandler Facsimile Editions in American Literature. San Francisco: Chandler.

———. [1892] 1980. *Pudd'nhead Wilson* and *Those Extraordinary Twins*. Reprint, 2 volumes in 1, edited by Sidney E. Berger. New York: W. W. Norton.

Evolution and Eugenics in American Literature and Culture, 1880–1940

I
Evolutionary Theory in American Literature

"It is the race instinct!": Evolution, Eugenics, and Racial Ambiguity in William Dean Howells's Fiction

Justin D. Edwards

RECENT HOWELLS CRITICISM HAS HIGHLIGHTED HIS INTEGRATIONIST views on American race relations. William L. Andrews, for instance, has shown the influence of Booker T. Washington's "reconciliationist" theories on Howells's position on racial difference (1976, 333): Howells believed that "reconciliation" of the black race "with the white race" could develop if African Americans could "learn at least provisional submission" (1900, 282). And Sarah B. Daugherty points to the racial doctrines of John Fiske—which attempted to show that human development was generated through the "peaceful intermarriage and natural selection" of races—as a source for Howells's ideas concerning racial mixture (1998, 57). But Elsa Nettles notes that while he agreed with many of Fiske's views on eugenics, evolution, and racial hierarchies, Howells "did not participate in the Immigration Restriction League," of which Fiske was the chief architect and a staunch supporter (1988, 102). I would suggest, then, that Howells's attitudes concerning racial difference cannot be reduced to an integrationist or a reconciliationist paradigm; Howells's writings on race and ethnicity are far too ambivalent and ambiguous to fit into a simple pattern or category. For me, the question is not whether Howells was a racist or an integrationist. Rather, it is how Howells's fiction is manipulated to move in both directions, fluctuating between progressive ideas on race and ethnicity, and reactionary ideas on racial difference. A good place to begin examining this fluctuation is by considering the relationship between Fiske's writing about eugenics and Howells's fiction.

Indeed, one of the models for Howells's fictional presentations of "racial science" comes directly from the racial doctrines of

John Fiske, an evolutionist who attempted to show that human development was generated through the "peaceful intermarriage [between whites and blacks] and natural selection" (1902, 77). Howells referred to Fiske's early enthusiasm for evolution in his review of Fiske's *Destiny of Man Viewed in Light of His Origin* (1884) and *Idea of God as Affected by Modern Knowledge* (1885). Howells was particulary interested in Fiske's integrationist theory because it contradicted the scientific views of the polygenesists, who argued that blackness and whiteness constituted discrete racial categories and that, by extension, Caucasians and Africans were two distinct species (Howells 1901, 732). By marking the constitutive difference of the races, early "racial scientists" such as John Campbell and Josiah Nott could claim that nonwhites were of a lower species and "not properly human" (Nott 1843, 255). For Campbell and Nott, such theories served as proof that interracial sexual activity was "unnatural," a crime against nature (Campbell 1851, 149). And contrary to the racial ideas of Fiske and Howells, Nott had claimed that the merger of two distinct species would lead the United States into cultural and physical degeneration (1843, 253). Thus, polygenesists rejected genetic integration by arguing in favor of racial purity, asserting that "racial hybridity" would engender genetic contamination and a sterile and sickly nation.

Fiske's evolutionary ideas appealed to Howells because they challenged polygenesis, leaving the door open for racial integration. Howells even goes as far as inserting Fiske's words into his 1892 novel *An Imperative Duty*, when Dr. Olney reassures Mrs. Meredith that racial mixture is a form of genetic progression because it will lead to "the permanent effacement of the inferior type" (1970, 27). Fiske's theory that intermarriage provided a path towards reconciliation rather than racial and genetic degeneration inspired Howells to write against 1880s scientific theories—expressed by scientists like Joseph LeConte—which argued that "the mixing of the two primary races produced an inferior breed which must eventually perish in the natural course of race struggle" (LeConte 1880, 100). By taking this position, Howells was able to advance liberal racial science and avoid the sentimental tradition of the "tragic mulatto" in which miscegenation produces calamity rather than unity (Berzon 1978, 93).

Furthermore, in *An Imperative Duty*, Dr. Olney's views on "the race problem" are informed by nineteeth-century monogenesist theories concerning racial difference. To Miss Aldgate's comment that "there seems something unnatural in the very idea" of intermarriage, Olney echoes Fiske's position by stating that "sooner or later our race must absorb with the colored race; and I believe it will obliterate not only its color, but its qualities. The tame man, the civilized man, is stronger than the wild man" (Howells 1970, 27). For Olney, following Fiske, such a step was a positive and inevitable part of evolution. And Howells's narrative continues along this path by contrasting Dr. Olney with the overtly racist Mrs. Bloomingdale, who exclaims that "negroes are an inferior race, and they could never associate with whites because they could never be intellectually equal to them" (43). Olney challenges Bloomingdale by putting forth integrationist ideologies concerning the race question:

> We [whites] might recognise them [African Americans] as fellow human beings in public, if we don't in private; but we ignore, if we don't repulse them at every point—from our business as well as our bosoms. Yes, it strikes one as very odd on getting home—very funny, very painful. You would think we might meet them on common ground before our common God—But we don't. (20)

Such comments are akin to Fiske's argument in favor of "man's integration" and his claim that "negroes can progress under the guidance of higher races . . . [for] they are somewhat more teachable than any brute animals" (1902, 53).

It is clear from this quotation that Fiske's racial theories included a paternalist tone that resonated with the colonial ideology of the "white man's burden." He saw blacks as animal-like, and he constantly infantilized African Americans, claiming that within the formula of "cosmic law" it was the Caucasian's responsibility to "look after the African" (48). Joel Williamson notes that in the decade of the 1880s, paternalism was a dominant racial ideology that was linked to integration: white liberals, including Howells, believed that African Americans needed to be "taught how to integrate into civilisation" (1984, 84). Dr. Olney also takes this position. Speaking of African Americans, Olney states: "I don't see why one shouldn't associate with them. There are terms a good deal short of the affection we lavish on dogs and horses that I fancy they might be very glad of" (Howells 1970, 20). This conflation of blacks with animals is followed by Olney's theory that racial mixture will elevate African Americans to the level of whites by erasing their animalistic "proclivities . . . toward evil" and teaching them "virtue" and "good tendencies" (27). For Olney, then, Africans are a primitive, animalistic, and childlike people who must be raised and taught to live in "civilisation."

Fiske's racial doctrines were clearly influenced by the discourses of imperialism put forth by Thomas Carlyle and others, suggesting that white skin provided access to the qualities of enterprise, organization, and civilization (Carlyle 1899, 350). Such ideologies can also be found in Fiske's texts on Darwinism. An early advocate of *The Origin of Species* (1859), Fiske attempted to link evolution to racial hierarchies by examining the brain sizes of various races. In his study of brain capacity, Fiske argued that "the difference in volume of the brain between the highest and lowest man is at least six times as great as the difference between the lowest man and the highest ape" (in Haller 1971, 133). These measurements provided a pseudoscientific basis for his imposition of racial hierarchies: the "Teutonic and Aryan races," Fiske claimed, were at the "top of the evolutionary scale" due to their "larger brain masses," whereas the "non-Aryan Hindu" and the "black African" were placed near the bottom of his scale (1902, 48–49). In *Duty*, Dr. Olney participates in a similar kind of comparative racial analysis in order to develop a hierarchical system based on racial difference. "The Irish are quicker-witted than we are," he says, "they're sympathetic and poetic far beyond us. But they let the high constructive faculties of the imagination work, when they ought to use a little attention and common-sense" (Howells 1970, 19). Olney thus uses racial classification to undermine his initial praise of the Irish: their quick wit and creativity are overshadowed by their lack of common sense. Like Fiske's brain measurements, then, Olney makes racial generalizations based on what he sees as mental capacities and intellectual abilities.

But Howells's *An Imperative Duty* is not the only text in which we can find his views on eugenics, evolution, and racial difference. In his autobiographical work, *A Boy's Town* (1890), for instance, Howells recounts his youthful abolitionist fervor and belief in "inter-racial brotherhood" alongside a secret that he has kept, one of the "ghosts of his youth" (129). This secret consists of an early memory wherein he became part of a "white mob" set against an African American rumored to have struck a white boy (129). Howells, who was raised in an abolitionist and radical Whig family, is haunted by his participation in this mob, and he reflects upon how boys and men can respond with their impulses when confronted with racial conflict. This early memory sets the stage for what Howells would later term "race instinct," the impulse that compelled him to join in a white mob and dismiss all of his enthusiasm for racial brotherhood.

Howells's movement in *A Boy's Town* between his liberal racial politics and his participation in the "white mob" paints a complex psychological portrait of his ambivalent responses to blackness. The complexity of his ambivalence is also present in *A Hazard of New Fortunes* (1890). Here, an ambiguous attitude toward African Americans is expressed by Isabel March when she and Basil meet the black janitor of an apartment they are visiting. On the one hand, this janitor inspires Isabel to celebrate the "angelic" aspects of the African American character, and she "los[es her] head" in his enchanting presence so that she momentarily falls "in love with the whole race" (36). On the other hand, though, Isabel's newfound love has a patronizing and stereotyping tone: she displaces the janitor's individuality by transforming him into a symbol of essentialized blackness and a model of his race. The janitor, from Isabel's perspective, lacks a personality, and she has no trouble conceiving of herself as a proud slaveholder: "If I had such a creature, nothing but death should part us, and I should no more think of giving him his *freedom*" (35). Isabel's stereotyping and condescension lead her to universalizing comments that call for a heavenly association between blacks and whites: "We shall be black in heaven," she says (36). Underlying Isabel's conception of a metaphysical, interracial brotherhood is the implication that any link between black and white culture is otherworldly. The joining together of the two races in the material world is not a possibility.

This resistance to racial integration is also seen in the mapping of New York streets in order to draw the boundaries between the city's various cultural communities. Basil's and Isabel's descriptions of New York draw lines to separate "us" from "them," even as the othered ethnic communities are being idealized and revered in the narrative voice:

> They [Isabel and Basil] had crossed Broadway and were walking over to Washington Square. . . . They met Italian faces, French faces, Spanish faces, as they strolled over the asphalt walks. . . . They met the familiar picturesque raggedness of Southern Europe with the old kindly illusion that somehow it existed for their appreciation, and that it found adequate compensation for poverty in this. (43)

Here, the ethnic diversity of the city is not disdained for its foreignness, but it is esteemed for its picturesque qualities and appreciated by the touristlike gazes of the Marches. Their "appreciation" of the "poverty" and "raggedness" of these Southern European immigrants not only transforms the tenement slums into what Amy Kaplan calls a "homely tableaux," but it also undermines the Marches' anxiety about American immigration policies and the ethnic diversity of New York (1988, 50). That is, the immigrants are separated from other Americans: they are confined within the tenement district, and the Marches must cross the borders of Broadway and Washington Square in order to glimpse the foreignness of this enclave. By setting up these borders, the narrative constructs a dividing line between the various ethnic communities that resists integration or social mixture. The Italian, Spanish and French immigrants, then, are forced into particular pockets of the city, constructing what Henry James would later call the "alien districts" of New York (1994, 95).

Howells thus expresses competing and contradictory ideas on racial and ethnic difference, which often move between a celebration of differentiation and the patronizing, essentialist representations of blacks, Italians, and Irishmen. While his positive depictions of ethnic and racial minorities can be traced back to his abolitionist roots, his more reactionary presentations of race and ethnicity are more difficult to explain. One possibility is that Howells was influenced by what Joel Williamson has referred to as the "1880s conservative reaction" to the persisting presence of Southern African Americans and Southern European immigrants in northern cities like Boston and New York (1984, 85). According to Williamson, the census of 1880 indicated that the increase in African American migration to Boston and an influx of Irish and Italian immigrants had significantly changed the city's demography. This population shift, Williamson argues, generated a conservatism whereby white liberals adopted reactionary positions concerning the "Negro problem" and American immigration policies (85). Although there is little evidence to suggest that Howells embraced an antiblack or anti-immigrant ideology at this time, his apprehension concerning the migration of African Americans would account for Howells's "perfectly stereotypical" and dehumanizing depiction of African Americans in *A Hazard of New Fortunes* (Wonham 1995, 709).

The Marches' mapping of the ethnic diversity of New York anticipates Dr. Olney's observations about Boston's changing population, the "Irish infiltration" of the city, in *An Imperative Duty* (1970, 12). Howells's novella begins with a description of Olney's walk through Boston's streets, which he sees for the first time after his "long sojourn in Europe" (4). Olney is surprised at how the city has changed; he is overwhelmed by Boston's "mixed humanity," the Irish, Italian, and African American citizens whom he, like James, refers to as the "aliens" of the city (6). He is disturbed by these "adoptive citizens" who surround him, people who are, in his words, "hardly American." For instance, he describes the Irish Bostonians as "old-world peasants" who are "thin and crooked with pale, pasty complexions" and other "physically delicate" features (4). As in *Hazard*, Howells's narrator questions American citizenry by asking what it means to be American in a country of immigrants and "aliens."

Furthermore, Olney is said to look at these people "scientifically," as if he is forming a medical opinion of them, of which his diagnosis is that these Irish "aliens" lack the handsome features of the "elder American race," for they are "coarse," "weak," "awkward," and "undeveloped" (4). The medical eye with which Olney examines the Irish community then turns to survey the American immigration policy that welcomes these "aliens"; after some contemplation, he concludes that the "transition from the Old World to the New, as represented in them [the Irish], was painful" (4). The pain of this transition frames their migration in physical and medical terms to suggest that a form of illness has infected the "physically delicate" constitution of the Irish citizens. But such comments also suggest that the Irish integration into the American body politic was also a painful development for the nation. Here, Howells implies that the naturalization of these "aliens" is unnatural, for it causes a pain that, from Olney's medical perspective, could infect the health of the national body.

The narrative's focus on Boston's Irish community is significant for introducing a novel about racial passing and the anxiety generated by the breakdown of ethnic and racial categories. For, as David Roediger points out, in 1890's America, the whiteness of the Irish was questioned and their appearance was often compared to the so-called "simian features" of African Americans (1991, 133). Theories about the "Irish race" abounded: some ethnologists argued that the "race" originated in Africa, while others claimed that Irishmen were neither black nor white yet both (135). Still other American racial theorists maintained the whiteness of Irishmen, arguing that they must be categorized as white in relation to the Native American and African American population of the United States (Dyer 1997, 53). In *Duty*, Dr. Olney articulates the racial confusion surrounding Irish identity:

> There is something very puzzling to us Teutons in the Celtic temperament. We don't know where to have an Irishman. We can predicate of a brother Teuton that this will please him, and that will vex him, but we can't of an Irishman. . . . They can't understand the simplest thing from us. . . . They seem more foreign to our intelligence, our way of thinking than the Jews—or the negroes even. (Howells 1970, 18–19)

Here, Howells's narrative brings up the question of an Irish racial identity alongside issues of class and social status: where does the Irishman belong in the social network of American society? Olney is unable to formulate a response to this question because, from his perspective, the Irishman is beyond understanding; there is a tremendous gap in communication and intelligence that separates the "Teutonic" from the "Celtic." This makes the Irish community not only "puzzling" and "foreign," but also illegible, for Olney is unable to read the Irishman's position in American racial hierarchies (they "seem more foreign . . . than the Jews—or the negroes even"). But Olney's thoughts here pose more than mere social questions; they also have "scientific" ramifications. That is, they echo the theories put forward by the racial scientist Nathaniel S. Schaler, who, in the 1890s, argued that American immigration policies should privilege the "Teutonic branch of the Ayran race" because its genetic constitution was better acclimatized to the North American environment (1896, 738). Irishmen, he stated, lay outside of this "Teutonic branch" and therefore the Celt's physical constitution was not given to "industrious" or hard work; because of this, he argued, Irishmen were too "lazy," "immoral," and "shiftless" to warrant access to American citizenship (738–39).

Howells's descriptions of the Irish community in *Duty* and the Southern European immigrants in *Hazard* illustrate the instability of whiteness as a racial indicator: his narratives question who is included in and excluded from the privileged category of whiteness. Such queries, as Richard Dyer points out, reveal differentiations within the very classification of whiteness, even among those whose racial identity is not in question (1997, 57). The Irishman, for instance, may very well have fair skin, but this does not necessarily give him access to white privilege. Dyer goes on to say that it is this "unstable, unbounded . . . category of white" that provides the strength of whiteness, for such fluidity "enables whiteness to be presented as an apparently attainable, flexible, varied category, while setting up an always movable criterion of inclusion" (57). While it is difficult to refute Dyer's claim, I would add that the flexible nature of whiteness, particularly in 1890s America, was also a source of anxiety. For during a historical period of increased immigration, the shifting category of whiteness forced people to question what American citizenship really meant. American identity, then, was anxiously placed under a microscope to determine who fit the mold of American citizenship and who did not. This determination was, of course, intimately linked to determining who was white. But because the borders of whiteness were unbounded, the question of American citizenry was difficult to define, thus creating an anxiety about what it meant to be American.

By introducing *Duty* with this discussion of the "Irish race," Howells sets in motion questions about racial difference: how do we categorize the Irish in terms of race? Olney asks, and what color are they, black or white? Structurally, these questions foreshadow the larger racial question of the text: what race is Rhoda? She is a woman who has lived her life as a white American, unaware of her African American heritage until it is revealed to her by her aunt, Mrs. Meredith, who fears that Rhoda's marriage to a white man will have tragic consequences if Rhoda gives birth to a dark-skinned child. Once this exchange between aunt and niece has taken place, Rhoda is faced with a profound identity crisis, for she transforms from "white" to "black," and Olney, who has fallen in love with her, must confront the realization that he loves a woman of "negro descent" (1970, 31).

Howells's ambivalent position on racial difference also arises in *Duty* when the narrative voice comments on Boston's African American community. "They all [had] barbaric taste in color," the narrator says of African Americans, "some of the young fellows were very effective dandies of the type we were beginning to call dude. . . . They had that air of being clothed through and through, as to the immortal spirit as well as the perishable body" (7). Here, Olney's descriptions highlight the bodies of the men and women who surround him, and his diction—"barbaric," "dandy," "dude"—trivializes them as predominantly physical rather than intellectual beings. Their intellectual presence is not mentioned, as if their identities are defined only through their bodies and clothing.

Although Olney argues in favor of racial integration, it is clear that he sees a large gulf separating the various racial and ethnic citizens of Boston. Such a gulf echoes the Fiskean doctrine of racial taxonomy, a doctrine that became popular during the 1880s and 1890s due to questions about who could emigrate to the United States. In fact, Fiske's theories got him elected president of the Immigration Restriction League in the 1890s, after he adopted the reactionary platform that argued in favor of "an enduring Anglo-Saxon birthright" in the United States (Haller 1971, 133). In *Duty*, Olney believes that this birthright has been forsaken; upon his return to Boston, he explains that "he ought, as an American, to live in America" but that due to American immigration policies he is "hopelessly alien" in the city of his youth (Howells 1970, 8, 9). These conservative remarks exist alongside his liberal position concerning intermarriage, thus highlighting Olney's conflicted views on racial difference. But perhaps his conservative and liberal attitudes are not as contradictory as they seem on the surface. That is, Olney's position on racial integration is based on the assumption that the population would benefit by gradually becoming whiter; the traits that he sees as native to blackness would eventually be bred out. White characteristics, he implies, are superior and therefore will remain dominant, destroying black traits. Thus, while he might hold a progressive position on miscegenation, he assumes that it is better to be white and that sexual reproduction is the key to achieving whiteness.

Theories about the possibility of achieving whiteness through racial mixture are also central to an exchange that takes place between Mrs. Meredith and Dr. Olney (26). Mrs. Meredith is concerned by the potential "persistence of ancestral traits" in Rhoda's children, and she asks Olney if he "believes in heredity" (26). When pressed to define what she means by this, Mrs. Meredith confesses that she is interested in atavism, the "transmission of character and tendency; the reappearance of [racial] types after several generations" (26). Olney's response is significant as he says,

> Take the reversion to the inferior race type in the child of parents of mixed blood—say a white with a mulatto or quadroon. . . . You hear of instances in which the parent of mixed race could not be known from a white person, and yet the child reverts to the negro type in color and feature. I should doubt it very much. . . . Because the chances are so enormously against it. The natural tendency is all the other way, to the permanent effacement of the inferior type. . . . The chances of atavism, or the reversion to the black great-great-great grandfather are so remote that they may be said hardly to exist at all. (26–27)

Here, Olney furthers his ambivalent racial doctrines. On the one hand, he does not dismiss miscegenation as immoral, but instead encourages racial mixture as central to the process of racial integration and human evolution. On the other hand, though, the language of taxonomy and racial hierarchies—words like "reversion," "permanent effacement," and "inferior type"—highlight his belief in white superiority. In fact, according to his argument, it is the very superiority of whiteness that makes the return of black ancestral features virtually impossible. For him, genetic atavism is a fiction, something that is an "effective bit of drama" for writers like George Washington Cable, but that this "reversion to the inferior race" holds no scientific currency (27).

Mrs. Meredith, however, remains unconvinced by Olney's argument. She wishes that Rhoda would marry an Italian, for this would enable her to continue the secret of her niece's birth. That is, Mrs. Meredith categorizes Italians as "dark" and thus any future "non-white" children from such a union would be explained by the child's Italian blood (35). The thought of Rhoda marrying a white American and conceiving a child, though, forces Mrs. Meredith to reveal the "stain upon that poor child's birth" (30). She admits to Olney that there is a "sinister tendency" in Rhoda's birth and an "ancestral infamy" that could manifest itself on the faces of Rhoda's children (30). Here, Mrs. Meredith's ideas about genetic atavism are very different from Dr. Olney's. For, in speaking of the return of blackness, she is filled with anxiety, calling genetic atavism a "sinister" (almost ghostly) presence that will return to life through the uncanny persistence of Rhoda's "ancestral infamy." "It is the race instinct!" she tells Olney, "It must assert itself sooner or later" (39).

This return of repressed blackness, or the emergence of that which has been previously rejected by consciousness, creates an anxiety that Mrs. Meredith cannot ignore. And it becomes her "imperative duty" to reveal that which has been repressed. In *Duty*, then, Rhoda's racial heritage takes on an essentialist quality that cannot be submerged or held at bay because it threatens the established order of things. The attempted repression of essentialist racial characteristics develops a cumulative energy in the narrative that demands its release and forces itself to the realm of visibility through the threat of atavism. Furthermore, this essentialist notion of racial difference is present in Mrs. Meredith's descriptions of Rhoda. She sees her niece as repressing her "ancestral infamy"; but, she says to Olney, "it might come out in a hundred ways. I can hear it in her voice at times—it's a *black* voice! I can see it in her looks! I can feel it in her character—so easy, so irresponsible, so fond of what is soft and pleasant! . . . It's her race *calling* her!" (37, Howells's italics). Such views on racial difference posit Rhoda's "blackness" as something that is under the surface of her skin, something that is sometimes pushed to the surface, threatening to release itself.

Mrs. Meredith's description of Rhoda, then, moves away from Fiske's integrationist theories. Instead, she echoes the racial theories of the zoologist and paleontologist Edward Cope, who during the 1890s reacted against the supporters of miscegenation and integration by warning against the so-called dangers of genetic atavism; "Its evils," he claimed, "[could] have innumerable ramifications throughout the body politic" (Cope 1890, 2400). The national body, he went on to say, was threatened by "the danger which flows from the presence of the negro . . . [and] the certainty of the contamination of the race" (2402). For Cope, the return of hidden racial traits was dangerous because the Negro mind—even if it were diluted with white blood—made him "unfit for American citizenship" (2401). Although Cope's position serves as a counterpoint to Howells's (and Olney's) theory of racial integration, we can see similar racial discourses in both positions. Both are concerned with the potential reappearance of that which is below the surface. Moreover, Cope, anticipating Howells, raises the question of racial categories alongside questions of who can be an American. People of mixed race, he claims, should not be allowed the privilege of American citizenship, for the threat of racial mixture could cause a deterioration in the moral, intellectual, and political fiber of the nation (2420).

While Howells does not go this far, he does set in motion questions about "whiteness" and its relationship to American citizenship. As a result, a national brand of "racial science" proliferates in his fiction at a time when issues of racial difference through African American migration and immigration are at the forefront of the national consciousness. This perceived time of crisis due to the "negro problem" and immigration forces a reevaluation of national identity with an urgency that is expressed in terms of race and ethnicity.

References

Andrews, William L. 1976. "William Dean Howells and Charles W. Chesnutt: Criticism and Race Fiction in the Age of Booker T. Washington." *American Literature* 48 (Fall): 327–39.

Banta, Martha. 1970. Introduction to *The Shadow of a Dream*. By William Dean Howells. Bloomington: Indiana University Press.

Berzon, Judith R. 1978. *Neither White nor Black: The Mulatto in American Fiction*. New York: New York University Press.

Campbell, John. 1851. *Negro-Mania: Being and Examination of the Falsely Assumed Equality of the Various Races of Men*. Philadelphia: Campbell and Power.

Carlyle, Thomas. [1849] 1899. "Occasional Discourse on the Nigger Question." In *Critical and Miscellaneous Essays* Vol. 4. London: Chapman and Hall.

Clymer, Jeffory A. 1998. "Race and the Protocol of American Citizenship in William Dean Howells' *An Imperative Duty*." *American Literary Realism* 30 no. 3 (Spring): 31–52.

Cooper, Anna Julia. [1892] 1988. *A Voice from the South*. Reprint, New York: Oxford University Press.

Cope, Edward. 1890. "The African in America." *Open Court* III (January): 2400–2421.

Daugherty, Sarah B. 1998. "*An Imperative Duty*: Howells and White Male Anxiety." *American Literary Realism* 30 no. 3 (Spring): 53–64.

Dyer, Richard. 1997. *White*. New York: Routledge.

Fiske, John. 1902. *Outlines of Cosmic Philosophy*. Vol. 4 of *John Fiske's Miscellaneous Writings*. Boston: Houghton Mifflin.

Fleming, Robert E. 1971. "Irony as a Key to Johnson's *Autobiography of an Ex-Colored Man*." *American Literature* 43 no. 1 (March):83–96.

Haller, John S. 1971. *Outcasts from Evolution: Scientific Attitudes of Racial Inferiority, 1859–1900*. New York: McGraw Hill.

Howells, William Dean. 1890. *A Boy's Town*. New York: Harper Brothers.

———. [1890] 1960. *A Hazard of New Fortunes*. New York: Bantam.

———. [1890] 1970. *The Shadow of a Dream*. In *The Shadow of a Dream* and *An Imperative Duty*, edited by Martha Banta. Bloomington: Indiana University Press.

———. [1892] 1970. *An Imperative Duty*. In *The Shadow of a Dream* and *An Imperative Duty*, edited by Martha Banta. Bloomington: Indiana University Press.

———. 1900. "An Exemplary Citizen." *North American Review* (August): 280–88.

———. 1901. "John Fiske." *Harper's Weekly* (20 July): 732.

James, Henry. [1907] 1994. *The American Scene*. New York: Penguin.

Kaplan, Amy. 1988. *The Social Construction of American Realism*. Chicago: University of Chicago Press.

LeConte, Joseph. 1880. "The Effect of Mixture of Races on Human Progress." *Berkeley Quarterly* 1 (April): 84–101.

Nettles, Elsa. 1988. *Language, Race, and Social Class in Howells's America*. Lexington: University Press of Kentucky.

Nott, Josiah. 1843. "The Mulatto a Hybrid—Probable Extermination of the Two Races if Whites and Blacks are Allowed to Intermarry." *American Journal of Medical Sciences* 6:252–56.

Roediger, David. 1991. *The Wages of Whiteness: Race and the Making of the American Working Class*. London: Verso.

Shaler, Nathaniel S. 1896. "Environment and Man in New England." *North American Review* (June): 723–45.

Williamson, Joel. 1984. *The Crucible of Race: Black-White Relations in the American South Since Emancipation*. New York: Oxford University Press.

Wonham, Henry B. 1995. "Writing Realism, Policing Consciousness: Howells and the Black Body." *American Literature* 67 no. 4 (December): 701–24.

His and Herland: Charlotte Perkins Gilman "Re-presents" Lester F. Ward

Cynthia J. Davis

CHARLOTTE PERKINS GILMAN ONCE WROTE THAT TRUTHFUL FICTIONS should "teach us life easily, swiftly, truly; *teach not by preaching but by truly re-presenting*; and we should grow up becoming acquainted with a far wider range of life in books than could ever be ours in person" (1970, 101; emphasis added). For Gilman, didacticism and mimesis were conjoined rather than antithetical goals. The fiction she produced in her lifetime exemplifies this conjunction. Even so literary an example as "The Yellow Wall-Paper" (1892) was written, she claimed, to correct the distorted views put forward by S. Weir Mitchell and other doctors ("Why I Wrote" 1980). Gilman sought to expose in "The Yellow Wall-Paper" the "truth" lurking behind medical falsehoods about women's nature. I want to suggest that her utopian novel *Herland* is no less informed by a desire to convey certain biological "truths" concerning female subjectivity. The difference resides in the way the novel represents reproduction as *parthogenetic* rather than, as does Mitchell, *pathogenic*. Indeed, among its other possible attributions, *Herland* deserves to be read as a response not to Mitchell but to another man of science, the so-called "Father of Sociology," Lester Frank Ward, whom Gilman once identified as the "greatest man" she ever knew (1972, 187). As I will demonstrate, *Herland* provided Gilman with a fictional vehicle through which to "truthfully re-present" Ward's evolutionary theories. My aim is not to present Gilman as Ward's mouthpiece—their views on gender and on eugenics alone distinguish them—but instead to establish the overlap between the two turn-of-the-century thinkers as grounds for the reevaluation and possible recuperation of both.

Gilman was aware of Ward several years before she met him. An article he published in 1888 on the female origin of the species was the piece of his that most fired her imagination, lending as it did scientific credence to many of her own reformist impulses.[1] In 1894, she twice made enthusiastic mention of Ward's theories in *The Impress,* the journal of the Pacific Coast Women's Press Association she was editing and struggling to keep afloat. When in 1896 Ward wrote to request a copy of her poetry collection *In This Our World* (he had particularly enjoyed her celebrated satirical poem "Similar Cases"), Gilman seized the opportunity to strike up a correspondence. The two cemented their friendship at a Woman's Suffrage Convention held in Washington, D. C. later that same year, where Ward held a reception in her honor. The man Gilman called "Professor" soon became one of her most influential mentors, and she would go on to promote her version of his views in lectures as well as in such venues as the *Woman's Journal* and her own one-woman journal, *The Forerunner.*

This is not to say that their friendship was unwavering. Gilman was clearly hurt that Ward apparently had not read any of her works, including her widely touted *Women and Economics* (1898), which she had dedicated to him (though he probably would have been discomfited by the intensity of its focus on both women and economics). She also objected to his publicly bemoaning the fact that no one else had taken up his theories in a systematic fashion, especially since she saw herself as working tirelessly to promote them.[2] This may explain why she would inform him that for all her admiration, "I have not really read you at all" (2 January 1907, Ward Papers). It may also explain why, after stressing her unfamiliarity with his work, she would stake a claim to her own ingenuity by expressing surprise and amusement at the "similarity of our point of view" (31 January 1907, Ward Papers). In a letter written the next day, however, she backtracked somewhat, assuring Ward that her *Human Work* "followed the same lines of thought you have covered so much more fully" (1 February 1907, Ward Papers). This string of letters seems to have prompted Ward to read *Human Work* three years after its publication, and his response was doubtless intended to flatter: "I could hear my own voice all the time. But, of course, it was not an echo. It is pitched much higher than I can strike and differs entirely in *timbre.* . . . all I could do was to block out the statue from the slab in rough strokes. . . . Now you come along and touch it up with a fine-pointed chisel" (9 February 1907, Gilman Papers). Gilman would continue to "chisel" away at his ideas in later works, including her 1911 *The Man-Made World,* which she also dedicated to Ward.

Though she publicly acknowledged her debt to the sociologist in several venues, it would be best to consider her as Ward's editor rather than disciple. Sandra M. Gilbert and Susan Gubar have pinpointed gender as one place where the two theorists diverge. In particular, they contend that "Ward's contrast between the race-type female and the sex-type male shares little with Gilman's. . . . Ward characterizes masculine 'appetition' with delight rather than distaste" (1999, 213). Gilman's distaste, by contrast, could not be more evident, while her celebration of woman's primacy was far more jubilant than Ward's. William Doyle puts their difference down to style: "Mrs. Gilman was a polemicist, one of that 'impolitic class of theoretical reformers' for whom Ward had predicted very difficult prospects, and whose actions he expected to 'engender a deplorable reaction'" (1960, 175). We might say that Ward provided the theory that Gilman then appropriated, enlarged, and politicized in order to lend "scientific" credence to her arguments for a radically restructured future and a liberated gender politics.

Although both their ends and their means may have diverged, Gilman and Ward shared a viewpoint she herself styled "similar," especially regarding Reform Darwinism. Ward had read his Spencer thoroughly and objected to the latter's laissez faire individualism, advocating instead a more hands-on, collective approach to social reform. Unlike Spencer, Ward came from a working-class background, abhorred class divisions, and knew firsthand that intervention and education could play key roles in determining future success. He was also critical of Spencer's reverence for natural competition, particularly as a model for human behavior, believing that it checked rather than enhanced evolution. Ward's reformist sympathies are most apparent in his rebuttals of Spencer, where he held that "the very law of evolution threatens to destroy hope and paralyze effort. Science applied to man becomes a gospel of inaction . . ." (1970, 20). Fearing inaction and the pessimism that prompts it, Ward did not jettison Darwinism so much as he reinterpreted it to support his rosier world view. As did many progressives, at least initially, he resisted biological determinism and embraced Lamarckianism, as the belief in the biological transmission of *acquired* traits helped explain variation and emphasized human agency. Further, while Spencer had focused attention on the "genetic" processes of evolution, Ward highlighted as well the "telic" (i.e., the conscious, human-engineered, "artificial") improvement of society, proclaiming the latter a more "evolved" form of evolution. As he wrote, "Social progress is either genetic or telic. . . . In the early human stages it is mainly genetic, but begins to be telic. In the latter stages it is chiefly telic. The transition from genetic to telic progress is wholly due and exactly proportionate to the development of the intellectual faculty" (1899, 179). The sole aim of the telic stages was happiness, according to Ward, and humans had it within their reach to achieve this goal because, unlike other species, they possessed rational intelligence or mind. A firm believer in evolutionary progress, Ward felt that it was best assured through conscious, collective human effort.[3]

Buoyed by Ward's theories, Gilman optimistically revised Darwin's evolutionary theory until it could support a unilinear narrative of progress. Viewing human life—both individual and collective—as an organic entity, Gilman regarded all of its forms, for so long as they were duly governed by natural laws, to be not only constantly evolving but clearly improving. And since biology dictates that "all the tendencies of a living organism are progressive in their development" (1966, 59), all that does not lend itself to the organism's development must be intrinsically alien to it. As Gilman insists in her preface to *Women and Economics*, "some of the worst evils under which we suffer, evils long supposed to be inherent and ineradicable in our natures, are but the result of certain arbitrary conditions of our own adoption, and . . . by removing these conditions, we may remove the evils resultant" (xxix). If nature were left to its own devices, both male and female would continue to flourish over time.[4]

To a greater degree than Ward, Gilman was concerned with why women had not flourished, finding her answer in the environmental forces that redirect, even inhibit what she considered inherent evolutionary tendencies. Despite our best efforts and natural impulses to progress, extrinsic factors including climate, habitat, nourishment, and other people both "form and limit" a given creature (1966, 4). The word "limit" is crucial to Gilman's brand of determinism: if it is "natural" for humans to evolve, then environmental circumstances typically operate as devolutionary, degenerative forces. For Gilman, the "natural" is not reduced to or replaced by the social: it is retained as a relatively plastic category that, for all its naturalness, is still capable of being altered through its functions, dysfunctions, and interactions.

While Gilman may have learned a good deal about Darwinism from Ward, it was Ward's original woman-centered theories about our origins that most excited her. In particular, his "Gynaecocentric Theory of Life" provided her with scientific justification to proclaim woman the "mother of the world" (1932, 331).

Gilman called this theory "the greatest single contribution to the world's thought since Evolution" (1972, 187) and "the most important contribution to the 'woman question' ever made." She wrote to Ward on 20 January 1904 that she meant to make his theory "the recognized basis of a new advance in the movement of women," and declared herself its principal advocate, preaching its truth at national and international conferences and in a score of articles (Ward Papers). She did not, however, buy his theory wholesale: in keeping with her perpetual optimism and intellectual competitiveness, she sought to expand upon his concept of the female origin of human life, declaring in a letter to Ward that the woman is "not only the original organism, not only equal [to man] in modification to species, but leader in modification to sex" (4 August 1904, Ward Papers). In other words, Gilman felt that women were improving more rapidly qua women than men were qua men, which is a more radical claim than Ward was ready to make.

As its etymology suggests, Ward's Gynaecocentric Theory posits the female as the race type, the necessary and primary force in evolution, and relegates the male to a secondary role based on his reproductive utility. For Ward, "woman is the unchanging trunk of the great genealogic tree; while man, with all his vaunted superiority, is but a branch, a grafted scion, as it were, whose acquired qualities die with the individual, while those of woman are handed on to futurity" (1888, 275). Life began, Ward contends, in a single fertile organism and that organism was female, leading him to conclude that "life was originally and essentially female" (1906, 542). He justifies this definitive gendering at the cellular level, where parthenogenesis is practiced, maintaining in "Our Better Halves" that "the asexual parent must be contemplated as, to all intents and purposes, maternal" and hence female (1888, 272). Since from the beginnings of life the only purpose of the male was to enable female reproduction, Ward derides the male as "insignificant," "an afterthought," "useless and a mere cumberer of the ground" (272). Such a theory was bound to appeal to Gilman as well as to other members of the woman's movement, especially when he gestured (if briefly and vaguely) toward its political resonance for the present and future: as Ward maintained, "it becomes clear that it must be from the steady advance of woman rather than from the uncertain fluctuations of man that the sure and solid progress of the future is to come" (275). Gilman clung to such tantalizing statements as she would to a life raft, finding validation therein for her own utopian leanings and inspiration for her exultation in "she who is to come" (the title of one of her poems).

In addition, Ward's female-centered theory allowed Gilman to give voice to her residual essentialist understandings of woman as mother without risk of pathologizing taint. Mothering remained a vexed topic for Gilman due to her oft-misunderstood "abandonment" of her own daughter[5]; Ward's theory provided her with scientific justification to talk about motherhood in terms that were not disconcertingly personal. Whatever her own failings as a mother, Gilman publicly subscribed to the belief that mothering was a woman's destiny. In an article entitled "Motherhood and the Modern Woman," for instance, she asks, "What is a woman?" then immediately answers, "The female of *genus homo*, one capable of being a mother" (n.d., 384). The slippage here that conflates gender identity and biological capacity informs Gilman's habitual glorification of "the maternal instinct, which goes so deep" (1892, 1). As she wrote in her 1898 *Women and Economics*, if motherhood were truly valued by the human race, "all its females [would be] segregated entirely to the uses of motherhood, consecrated, set apart, specially developed, spending every power of their nature on the service of their children" (19).

It is precisely this consecrated, gynocentric segregation that her 1915 novel *Herland* would depict. Gilman's ongoing correspondence with Ward confirms that she began envisioning her utopia several years before setting it down in fictional form. In the decade prior to *Herland's* publication, Gilman was already speculating about what might have happened to cause the overthrow of the original "matriarchate." A lengthy passage from a 1908 letter might even constitute a rough outline for the novel she would serialize in *The Forerunner* some seven years later. As she wrote to Ward, speculating about how this original matriarchy could have been displaced via evolutionary processes,

> Assume a matriarchal settlement under exceptionally good conditions—good climate, abundant food—peace & plenty. Assume an excess of females. (Under the difficult conditions of previous human life there had been an excess of males, with combat for selection by females, as in other animals.) Having now good condition and surplus females, the male becomes increasingly valuable. The dominant females, already the industrial power and used to tribal communion, now establish *a voluntary polygamy* agreeing to maintain one male to each small group of females. If this were done, the male, being now

> supported by the group of females and held in high esteem, is in a position to develop naturally [through] indulgence [his] cruelty, pride, etc. which would so lead to the more injurious effects of unchecked masculine rule. . . . Androcracy—the world as we have it. . . . It has always been a puzzle to me to see how the female, the unquestioned dominator of all previous life, could have been suddenly—or gradually—overthrown." (9 May 1908, Ward Papers)

The final sentence demonstrates Gilman's inclination to push Ward beyond his comfort zone, finding answers to questions Ward did not raise or care to probe. Though *Herland* does not depict the overthrow of a matriarchate, it does allow for its "gradual" possibility by introducing men, including an excessively "androcratic" one; moreover, together with its sequel, *With Her in Ourland* (1916), it simultaneously heralds women's dominance while suggesting how dispensable and manifestly cruel not only individual men but androcratic societies could be. Ward's gynaecocentric theory helped Gilman to explain present-day inequities even as it provided a potential, if unfinished, model for a utopian future. In short, it lent scientific stamp to feminist longing.

Indeed, it is possible that Ward may have provided Gilman with the impetus to venture into the field of utopian fiction. In a 1906 article by Ward entitled "The Past and Future of the Sexes" and found among Gilman's papers, the sociologist criticizes the recent boom in utopian fiction for its meager prophetic powers. Opposing mere speculation, Ward contends that "the only possible scientific basis for forecasting the future of the sexes is a study of their past history from the very origin of sex" (542). It was precisely this recommended "cosmological perspective" that Gilman deployed to yield the utopian vision she began sketching in print around 1907 and most clearly encapsulated in her fictional *Herland*.[6]

Ward's Gynaecocentrism might thus be said to provide the seed that blooms in *Herland*.[7] In this fictional female world, motherhood is not only the origin but also the endpoint of racial evolution. The women of Herland have established a society based upon both the principle and the practice of motherhood. As one of the guides explains to the male visitors, "Here we have Human Motherhood—in full working use . . . the children in this country are the one center and focus of all our thoughts. Every step of our advance is always considered in its effect on them—on the race. You see, we are *Mothers*" (1979, 66). A female race set apart, the Herlanders wage "no wars" and have advanced "not by competition, but by united action" (60). Like both Ward and Gilman, the Herland women have studied other species and recognized the inutility of the male among insects, birds, and cats (48). Disencumbered of masculine uselessness, they have created a world of useful mothers, a cooperative nurturing place where, if men have a role, it is only because they can make conception potentially easier (after all, parthenogenesis is no roll in the hay). Two of the three men also serve as models for Gilman's readers by overcoming their initial resistance and recognizing the superiority of Herland's matriarchy over "Ourland's" patriarchy. As Van—the sociologist-narrator and most open-minded of the trio—acknowledges, a Herland mother is a vastly different, vastly improved version of her sister in our land:

> They were mothers, not in our sense of helpless involuntary fecundity, forced to fill and overfill the land, and then see the children suffer, sin, and die, fighting horribly with one another; but in the sense of Conscious Makers of People. Mother-love with them was not a brute passion, a mere "instinct," a wholly personal feeling; it was a religion. It included that limitless feeling of sisterhood, that wide unity in service . . . and it was National, Racial, Human. (69)

Van's description of Herlanders as "Conscious Makers of People" gestures not only toward their parthenogenetic capabilities but also to the careful thought they have given to childbearing and child rearing. In both respects, Gilman's utopia bears the imprint of Ward, as it realizes not only his original matriarchal system but his belief in advancement through conscious or telic processes. For the Herlanders, conception is conceptual: a child begins as a thought in a woman's mind rather than when a sperm meets an ovum. Amazed at the Herlanders' "reasonableness"—especially when compared to the alleged irrationality of Ourland women—Van declares the "most astonishing" trait of the Herland women to be their "conscious effort to make it better" (77). When things needed changing, the Herlanders "sat down in council together and thought it out. Very clear strong thinkers they were" (68).

Even Gilman's mode of writing, didactic and at times pedantic, evokes this emphasis on intellect. After all, the utopian generic formula constructs readers who engage primarily with their minds rather than their hearts—they are successful when readers become willing to entertain the thought of such a world's possibility and desirability compared to their own. A utopia is an ideal, a good place that exists no place, and Herland would fit this formula—were it not for the fact that its creator, Gilman, was firmly convinced of such a world's prior existence, on the cellular level, as Ward's gynaecocentric model had proposed. And while it might seem that a utopia premised upon mother-worship was more reactionary than visionary—replicating as exemplary the very ideologies many feminist reformers, including Gilman, were seeking to dismantle—in Gilman's utopia maternalism is reconstrued as a progressive rather than regressive force. Or rather, it *is* regressive, literally speaking, to the extent that it evokes Ward's original matriarchate, but this regression is the sine qua non of Herland's glorious progress.

In a world this devoted to mothering, it is little wonder that the process of converting the three male visitors to the Herland maternal "religion" evokes the processes of gestation. The three Americans are confined within this female yet strangely unfeminine place for approximately "nine months"—a period Van refers to outright as "our confinement" (58). At the end of this gestational period, the insensitive Terry may be no closer to conversion, but the always chivalrous Jeff is fully reborn. Van's reaction to the matriarchy is more thoughtful than that of his friends:

> It gave me a queer feeling, way down deep, as the stirring of some ancient dim prehistoric consciousness. . . . It was like—coming home to mother. . . . I mean the feeling that a very little child would have, who had been lost—for ever so long. It was a sense of getting home, of being clean and rested; of safety and yet freedom; of love that was always there, warm like sunshine in May, not hot like a stove or a featherbed—a love that didn't irritate or smother. (142)

A sociologist recognizing not only a primary mother but the comfort and promise of a notion of maternal origins: In *Herland*, Gilman may not only be drawing on Ward but depicting him.

Regrettably, Ward's Gynaecocentrism is not the only "ism" informing this utopia. As Thomas Peyser (1992) first noted, Herland is a world in which racism and imperialism are underlying assumptions. The novel asserts the superiority (denoted by both altitude and physical traits) of the fair and pure Herlanders over the primitive "savages" who first introduce the Americans to this mythic land and who, despite their obvious foreknowledge, are stripped of their claims to discovery by the three Americans' imperialistic presumption that they are the first to encounter Herland. As a result, as Peyser concludes, Gilman reinscribes the dominant social order even while trying to reform it.

Gilman's belief in Anglo-Saxon superiority and the threat of amalgamation is not restricted to this novel but runs throughout her larger corpus, most notoriously in her published diatribes against the "yellow peril."[8] She was an advocate of "euthanasia for incurables" and she also endorsed compulsory sterilization for the "unfit" (1932). In her later years, Gilman became a decided proponent of eugenics and an advocate of nativist views. Judith Allen notes that Gilman moved from New York City to Norwich Town, Connecticut, in 1922 in part because of her distaste for the city's increasing ethnic diversity, and she and her husband soon became involved in the Norwich Americanization Institute (1997, 25 n. 67). Around this same time, Gilman began publishing a number of articles in major periodicals addressing the perils of immigration, asserting in one of these that "a social group of naked, dirty, witch-ridden savages is distinctly not 'equal' in any sense to a social group of clean, healthy, intelligent persons, but is markedly inferior in that which measures human values—social evolution" (1929, 419)—a hierarchy *Herland* also inscribes.

Lisa Ganobcsik-Williams has argued that Gilman's biases against immigrants and blacks actually stem "from her total commitment to the idea of human progress through social evolution" (1999, 17). In other words, the faith Ward helped to flame—i.e., the belief that evolution was not an utterly determined process but an agentive and uneven one—led Gilman to many of the views present-day readers find so abhorrent. The fact that these views owe much to evolutionary theories is revealed in Gilman's 1923 essay "Is America Too Hospitable?" where she maintains that "It is . . . a mistake to suppose that social evolution requires the even march of all races to the same goal. . . . Evolution selects, and social evolution follows the same law. If you are trying to improve corn you do not wait to bring all the weeds in the garden to the corn level before going on" (1923, 1985–86).

Gilman's racism is strongly conveyed in this article as well, as when she proclaims that "We are perfectly familiar in this country with the various blends of black and white, and the wisest of both races prefer the pure stock" (1923, 1985). Though it would be convenient to attribute Gilman's reactionary views to a conservative old age, they surfaced as well in earlier works, including *Concerning Children* (1900), where she refers to black children as "pickaninnies." Moreover, in a 1908 article entitled

"A Suggestion on the Negro Problem," she posits American blacks as more "primitive" than American whites and advocates for the creation of an "army" of the former who would provide labor in exchange for "civilizing" lessons (180–83).[9]

This support of eugenics and segregation is one point where Gilman and Ward diverge. If Ward is mentioned today, it is more often for his views on racial assimilation and eugenics than for his gynaecocentric theory. For instance, I recently came across Ward's name while preparing for a class on Charles Chesnutt's *Marrow of Tradition*. Reading up on context for the novel, I found Ward identified by Eric Sundquist as a principal turn-of-the-century advocate of the belief that blacks "naturally" want to rape or mate with whites to improve their own race.[10] There is truth to Sundquist's reading of Ward, though out of context it is rather unfair to Ward, who saw himself—and was generally seen by blacks and whites alike—as enlightened on racial issues. He sided with the abolitionists from early on and later became an outspoken foe of the eugenics movement and of prevailing beliefs in hereditary differences. Ward considered fallacious the widely-held beliefs in not only racial superiority but also racial uniqueness or purity. In his view, there were no valid, scientifically provable racial differences, just as there were no inferior races, only undeveloped or stunted ones. As he said of American blacks, "the only wonder is that they have not sunk lower into barbarism and degradation; and the very fact that they have not, argues well for their inborn talents and their natural and moral endowments" (1935, 263).

According to Ward, races were superior only by virtue of environment; biological differences, where and when they exist, were both insignificant and surmountable. The best way to surmount them, Ward held—and here is where he skates on dangerous ground, though it is also where he most clearly differs from Gilman—was through assimilation. As Ward put it, "there is no race or class of human beings who are incapable of assimilating the social achievement of mankind and of profitably employing the social heritage" (1906, 110). The evolution of man has resulted from differing races coming into contact and commingling, whether out of economic necessity, sexual attraction, and/or mutual needs for defense against enemies. Different races had mixed so often throughout history that no race could be considered pure, which was a good thing, since it was through assimilation that the human race advanced (1970, 206–11). While not without problems, in his racially divisive day such views were quite progressive, far more so than Gilman's more troublesome, though more mainstream, stance.

Ward also studied eugenics and found unscientific its basic premise—that breeding was the best way of bettering the human race—as it took no account of environment's strong impact in shaping human character. Ward argued that change was far more likely to ensue by altering the environment to allow hereditary forces to flourish rather than by relying on hereditary forces operating in an unchanged and frequently oppressive environment. If and when the environment could be perfected, he held, then and only then might it be worthwhile to experiment with eugenics (see Chugerman 1939, 422–26).

Yet even while Ward comes across as more liberal in this arena, he did condone negative eugenics as a means of breeding out the "unfit," and it is precisely a negative eugenics that the "Aryan" Herlanders practice (Gilman 1979, 54). They are, after all, dedicated from the beginning to the eugenicist goal of making "the best kind of people" (59), which entails regularly "breeding out" substandard mothers by encouraging the "worst types" not to reproduce (82). Such prophylactic measures were to Gilman so commonsensical and so efficacious in warding off the twin dangers of "idiocy and degeneracy" that she could blithely assert, in an article published the year prior to *Herland,* that "negative eugenics, . . . the legal sterilization of the unfit . . . is now accepted as wise and just in many communities. . . . That society has a right to thus arrest its own decay is questioned only by a few extreme individualists" (1914, 219). Clearly, for both Ward and Gilman, there was a fine line between a reformist's zeal for the evolutionary potential of the human race and a eugenicist's promotion of practices designed to realize "perfection" that much sooner.

Clearly, Gilman, for better or for worse, was never fully wedded to Ward's views across their spectrum. What she did wholeheartedly embrace was his notion of the self-sufficient female organism, capable of sustaining all life and of explaining its origins and possibilities. One explanation for the attraction, or at least one of its outcomes, is that Ward's model of the parthogenetic, essentially female original ovum enabled Gilman's desired confluence of mimetic and didactic modes: Gilman's fictional representation of Ward's model both reflects and teaches about a "prior" reality. In other words, despite utopian fiction's futuristic emphasis, *Herland* could—and I would suggest, should—be read as exemplifying Gilman's *mimetic* didacticism. After all, it forges a symmetry between the utopian future perfect and the biologized past reflected, if not fully realized, in Ward's gynaecocentric evolutionary account.

Notes

1. The article in question was "Our Better Halves" (1888). See Hill (1980) for further biographical information concerning Ward and Gilman, especially 263–71. Kessler (1995) discusses Ward briefly (32, 35–37) and suggests without exploring the issue that "Herland . . . is a thought-experiment extrapolated from Ward's theory" (37). Gilman confessed to having difficulty reading both Ward's *Dynamic Sociology* (1883) and *Pure Sociology* (1903)—she eventually reviewed the latter in *The Forerunner* in 1910. But it was primarily this one 1888 article that introduced her to Ward's ideas. See Hill (1980), Allen (1997, 133–34), and Magner (1992, 123–24) for useful overviews of Ward's influence on Gilman.

2. Ward did praise Gilman's efforts privately, as in this 11 February 1911 letter: "No one is doing as much as you to propagate the truth about the sexes, as I have tried to set it forth" (Gilman Papers). His praise here seems to be backhanded, however, since he represents Gilman as little more than his foot soldier.

3. Both Chugerman (1939) and Scott (1976) provide helpful background information on Ward's evolutionary views.

4. See also Magner (1992) on Gilman's relationship to Darwinism. In *His Religion and Hers*, Gilman does identify one example of a "natural beginning of an unnatural relation" (1976, 195) stemming from the Darwinian concept of sexual selection. Due to "the prolongation of infancy," due to the fact that humanity is a collective noun and individual interests are often subjugated for the good of the group, an inequity, a sort of temporal lag, gradually developed in the evolution of females versus males. Although this discrepancy pertains across all species, Gilman contends that it is among humans alone that this service to the child unnaturally evolved into service to the adult male as well, a practice soon naturalized as custom (195–217). The result, as she famously concluded some twenty-five years earlier in *Women and Economics* (1898), was that human beings became "the only animal species in which the sex-relation is also an economic relation" (5).

5. Just prior to the marriage of her first husband Walter Stetson to her dear friend Grace Channing, Gilman—then living in California and struggling to make ends meet—sent her daughter Katharine east to live with the couple, prompting a national outcry.

6. In this same article, "*The Past and Future of the Sexes*" (Gilman Papers), Ward acknowledges—though he does not explicitly name—Gilman by suggesting that "the only person who, to my knowledge, has clearly brought out this cosmological perspective, not merely in things human, but in the vast reaches of organic evolution, is a woman" (541). It is clear that he is referring

to Gilman by his subsequent references to her famous poem, "Similar Cases." Gilman began serializing in *The Times Magazine* a work entitled "A Woman's Utopia," from January through March of 1907, but left it uncompleted. Carol Farley Kessler identifies her first published complete utopia as "Aunt Mary's Pie Plant," which appeared in the *Woman's Home Companion* in June of 1908 (1995, 117).

7. A short synopsis of *Herland*'s plot might be useful for those who have not read the novel. Three male Americans, all friends and scientists, join a scientific expedition to a remote and "savage" country, where they hear legend of a land peopled only by women. Upon returning at a later date, they do indeed discover such a place, filled with creatures who, while female, bear little resemblance to the women of our own land, and who, without men, have been able to reproduce themselves parthogenetically. Imprisoned when they threaten to tell the world about the country they name Herland, the Americans are nonetheless treated with kindness. They are also pumped for information about their "bisexual" society, forced to draw comparisons that bathe their own country in a distinctly unfavorable light. And yet the Herlanders persist in believing that a world with two sexes must surpass their own; they think of themselves as "only half a people" (1979, 95) and are convinced of "the superiority of a world with men in it" (135). Thus, there is widespread rejoicing when the Americans eventually marry their Herland girlfriends. The honeymoons have barely commenced, however, when the macho Terry, fed up with his new bride's typically Herlandesque asexuality, tries to rape her, whereupon he is promptly and permanently expelled. The third American, Jeff, opts to stay in Herland, contented with his new-found paradise, but Van (the narrator and most likeable of the Americans) and his wife, Ellador, agree to accompany Terry back to the states (thereby setting the stage for the sequel, *With Her in Ourland*). A portion of my comments about *Herland* and Ward is derived from the fourth chapter of my *Bodily and Narrative Forms* (2000).

8. See Lanser (1989) for a discussion of Gilman's anti-Asian bias and its traces in "The Yellow Wall-Paper."

9. For more on Gilman's racism and eugenicist views, see Ganobcsik-Williams (1999) and Scharnhorst (2000) as well as Allen's unpublished essay (1997).

10. See Sundquist, who contends that Ward "held that the evolutionary character of racial conflict by which states and nations progressed naturally produced a corollary law that drove black men to rape white women" (1993, 410).

References

Allen, Judith. 1997. "The *Late* Gilman: Sexuality, Birth Control and Eugenics, 1911–1932." Unpublished paper read at the Second International Charlotte Perkins Gilman Conference. Skidmore College, Saratoga Springs, New York, June 26–28, .

Chugerman, Samuel. 1939. *Lester F. Ward: The American Apostle*. Durham, N.C.: Duke University Press.

Davis, Cynthia J. 2000. *Bodily and Narrative Forms: The Influence of Medicine on American Literature, 1845–1915*. Stanford, Calif.: Stanford University Press.

Doyle, William Theodore. 1960. "Charlotte Perkins Gilman and the Cycle of Feminist Reform." Ph.D. diss., University of California.

Ganobcsik-Williams, Lisa. 1999. "The Intellectualism of Charlotte Perkins Gilman: Evolutionary Perspectives on Race, Ethnicity, and Class." In *Charlotte Perkins Gilman: Optimist Reformer*, edited by Jill Rudd and Val Gough. Iowa City: University of Iowa Press.

Gilbert, Sandra M., and Susan Gubar. 1999. "'Fecundate! Discriminate!' Charlotte Perkins Gilman and the Theologizing of Maternity." In *Charlotte Perkins Gilman: Optimist Reformer*, edited by Jill Rudd and Val Gough. Iowa City: University of Iowa Press.

Gilman, Charlotte Perkins (Stetson). 1892. "Human Nature." *Weekly Nationalist* (1892). Charlotte Perkins Gilman Papers, Schlesinger Library, Radcliffe College, Cambridge, Mass.

———. 1908. "A Suggestion on the Negro Problem." *American Journal of Sociology* (14 July): 78–85.

———. 1914. "What May We Expect of Eugenics?" *Physical Culture* 31 (March): 219–22.

———. 1923. "Is America Too Hospitable?" *Forum* 70 (October): 1983–89.

———. 1929. "Unity is Not Equality." *World Unity* 4 (September): 418–20.

———. 1932. "Birth Control, Religion, and the Unfit." *Nation* (27 January): 108–9.

———. [1898] 1966. *Women and Economics: A Study of the Economic Relation Between Men and Women as a Factor in Social Evolution*. Reprint, edited by Carl N. Degler. New York: Harper Torchbook.

———. [1911] 1970. *The Man-Made World, or Our Androcentric Culture*. Reprint, New York: Source.

———. [1935] 1972. *The Living of Charlotte Perkins Gilman: An Autobiography*. Reprint, New York: Arno.

———. [1923] 1976. *His Religion and Hers: A Study of the Faith of Our Fathers and the Work of Our Mothers*. Reprint, Westport, Conn.: Hyperion.

———. [1913] 1979. *Herland*. Reprint, New York: Pantheon.

———. [1913] 1980. "Why I Wrote 'The Yellow Wallpaper.'" Reprint, in *The Charlotte Perkins Gilman Reader*, edited by Ann J. Lane. New York: Pantheon, 19–20.

———. [1900] 1990. *Concerning Children*. Reprint, Boston: Small, Maynard.

———. Charlotte Perkins Gilman Papers, Schlesinger Library, Radcliff College, Cambridge, Mass.

———. n.d. "Motherhood and The Modern Woman." *Physical Culture* 382–85. Charlotte Perkins Gilman Papers, Schlesinger Library, Radcliffe College, Cambridge, Mass.

Hill, Mary A. 1980. *The Making of a Radical Feminist, 1860–1896*. Philadelphia, Pa.: Temple University Press.

Kessler, Carol Farley. 1995. *Charlotte Perkins Gilman: Her Progress Toward Utopia With Selected Writings*. Syracuse, N.Y.: Syracuse University Press.

Lanser, Susan S. 1989. "Feminist Criticism, 'The Yellow Wallpaper,' and the Politics of Color in America." *Feminist Studies* 15: 415–41.

Magner, Lois N. 1992. "Darwinism and the Woman Question: The Evolving

Views of Charlotte Perkins Gilman." In *Critical Essays on Charlotte Perkins Gilman*, edited by Joanne B. Karpinski. New York: G. K. Hall.

Peyser, Thomas Galt. 1992. "Reproducing Utopia: Charlotte Perkins Gilman and *Herland*." *Studies in American Fiction* 20 (Spring): 1–16.

Scharnhorst, Gary. 2000. "Historicizing Gilman: A Bibliographer's View." *The Mixed Legacy of Charlotte Perkins Gilman*, edited by Catherine J. Golden and Joanna Schneider Zangrando. Newark: University of Delaware Press.

Scott, Clifford H. 1976. *Lester Frank Ward*. Boston: Twayne.

Sundquist, Eric J. 1993. *To Wake the Nations: Race in the Making of American Literature*. Cambridge, Mass.: Harvard University Press.

Ward, Lester F. 1888. "Our Better Halves." *Forum* 6 (November): 266–75.

———. 1899. *Outlines of Sociology*. New York: Macmillan.

———. 1906. *Applied Sociology: A Treatise on the Conscious Improvement of Society by Society*. Boston: Ginn.

———. 1906. "The Past and Future of the Sexes." *The Independent* (8 March): 541–45. Charlotte Perkins Gilman Papers, Schlesinger Library, Radcliffe College, Cambridge, Mass.

———. 1935. Young Ward's Diary, edited by Bernhard J. Stern. New York: Putnam.

———. [1903] 1970. *Pure Sociology: A Treatise on the Origin and Spontaneous Development of Society*. Reprint, New York: Augustus M. Kelley.

———. Lester F. Ward Papers. Brown University Library, Providence, R.I.

Jack London's Evolutionary Hierarchies: Dogs, Wolves, and Men

Lisa Hopkins

WHEN JACK LONDON PUBLISHED *THE CALL OF THE WILD* IN 1903, AND followed it up with its quasi-sequel *White Fang* in 1906, he was touching directly on a theme that was and has remained highly controversial in American culture: evolution. Behind these apparently innocent, heroic, and quasi Aesopian fables, with their mythopoeic nomenclature, lie profound concerns with changes in animal nature over time. Moreover, London's intricate interweaving of the stories of Buck and White Fang with those of their human masters does more than demonstrate his firm belief in environmental determinism; it also suggests that what is true of animals may be so of humans too. This connection was made early in the reception of the novels, and the developmental extremes of which Buck and White Fang ultimately prove capable led to London being often criticized (most notably perhaps by Theodore Roosevelt) for anthropomorphizing and sentimentalizing his canine creations. Though London certainly does draw parallels between animal and human behavior—as one would indeed expect from one committed to evolutionary theory—what seems to me much more striking are in fact the *differences* between his dogs and his men and what these may imply about London's understanding and use of the vexed differentiations between species and race.

Though both Buck and White Fang change dramatically during the course of their respective narratives, the human beings with whom they have to deal remain remarkably constant. John Thornton in *The Call of the Wild* and Weedon Scott in *White Fang* are always essentially good and deserving; Jim Hall, although actually innocent of the crime for which Judge Scott sentenced him, was nevertheless a bad lot in general, as his two previous convictions showed, and has not improved in prison. Contrary to Judge Scott's predictions, you *can* cure a chicken-killing dog, but human beings seem inherently recidivist. Similarly, Leclère in the short story "Bâtard" starts vicious and remains so. Indeed London makes much play of dogs' alleged ability to read human character instinctively and unerringly: White Fang may not know his own mother again, but although he has never heard of Jim Hall and knows nothing of his reasons for seeking revenge on Judge Scott, he immediately recognizes his felonious intent. London's dogs may learn and change, but his men, it seems, are what they are.

To some extent the reason is, as London constantly reminds us, that there are always two modes of being available to dogs. Humans may, as we now know, share 98 percent of their genes with chimpanzees, but we do not form part of the same breeding group; speciation has definitively and apparently irrevocably taken place. Dogs, by contrast, remain much closer to wolves and, crucially, can still interbreed with them. Indeed, White Fang is the product of precisely such a mating and is ultimately acclaimed by the Scotts as a "Blessed Wolf." So incomplete does the process of speciation seem to be that an individual animal is capable of effectively fast-tracking in its own person the entire history of the process.

This echoes the popular mid-nineteenth century idea that the development of the human embryo recapitulated that of the entire species, progressing from an initially reptilian form to an ultimately mammalian one. Adult humans, however, lack this fluidity. Though the discovery of Neanderthal man in 1856 had made it plain that there had once been species of hominids other than *Homo sapiens*, none survived (and the extent to which interbreeding had ever been possible continues to be debated). Man could not, in his own person, ever change his behavior so markedly as to find himself reclassified from one species to another as do both Buck and White Fang.

Even for dogs, however, the process is not fully two-way. After the death of Jack Thornton, Buck answers the call of the wild finally and irrevocably, not only running with the wild brother but actually heading the wolf pack. White Fang, however, is ultimately praised by the Scotts precisely for his retention of the characteristics of a wolf despite his assimilation into the family and the law. Although development in both directions is equally possible, it is clear that one has greater cachet, and is certainly better calculated to appeal to a writer like London, who had made his reputation through his own reputed closeness to the wild, as manifested in his oyster piracy and his Klondike and

"Northland" trips. Indeed perhaps London's ultimate evolutionary fantasy would be a human being who could answer the call of the wild as effectively as a dog.

And this is, I think, what he did in fact fantasize in *Before Adam,* the work that followed *White Fang.* Like *White Fang, Before Adam* was written in little more than a month, having clearly caught its writer's imagination. Like *White Fang,* too, it was initially serialized, allowing for an episodic narrative method that essentially boils down to construction-by-cliffhanger. There are also significant differences, though, and one of the most striking is that unlike both *White Fang* and *The Call of the Wild, Before Adam* is set in the past, not the present. This shift inherently qualifies and indeed arguably even vitiates the entire issue of progression—in whichever direction—that lies at the thematic core of the dog books. But if London has set himself one problem, he has solved another. Thus, I want to argue that here, as in the Northland stories, he has found in race a classeme for men that he finds equivalent to that of speciation in dogs, and that it is indeed the sensitivities inevitably attendant on race issues that complicate his treatment of progression. (I use the word "men" advisedly, for though the exchange of Indian women facilitates the developmental trajectory of London's white heroes, the women themselves remain static counters in the transaction.)

London's theories of race were complex. Though he is notorious for his alleged remark "What the devil! I am first of all a white man and only then a Socialist" (Foner 1947, 59), Jonathan Auerbach points out that miscegenation and flexibility of identity were both crucial parts of his ideas about race (Auerbach 1996, 58). Maxwell Geismar similarly argues that "the vicious tone of London's theories of racial supremacy applied mainly, as it were, to *poor* Negroes—Negroes or the other 'inferior races' who were inferior precisely because they had not as yet acquired breeding, wealth or social tradition" (Geismar 1954, 214). For London, racial classifications are in flux, and not only can reclassification be in some sense earned, but contact with the nonwhite also identifies and indeed defines the values of whiteness. Thus though London's men may not be able to choose between being dog or wolf, they can move between modes of being. London instantiates this condition as a reified and mythologized redness and whiteness with the Indian not only standing to the white man as the wolf does to the dog, but also, and more importantly for London's purposes, facilitating the transition of white man to the condition of the wolf.

London's representations of race and ethnicity are conditioned by more than his theories about Native Americans, however. The "Kipling of the Klondike" seems also to have been very aware of the extent to which he was negotiating literary territory explored by British rather than American writers. The three authors he is known to have read in the Klondike—Darwin, Milton, and Kipling—not only dealt in various ways with the question of human origins and nature, but also either participated in or provided crucial ideological ammunition for the British imperial adventure. London, therefore, does not only chart the difference between the evolution of men and dogs, he also negotiates the related ideological terrain of the evolution of ideas about evolution, and his own difference from the imperially-inflected models of his British comparators. Here, his commitment to the influence of environment does double duty because he is able to suggest that locale—in particular the Northland—plays a crucial part in facilitating a less deterministic and teleologically-oriented model than that traditionally found in British accounts of evolution.

London himself called *The Call of the Wild* a portrait of "devolution or decivilization of a dog" and *White Fang* "the evolution, the civilization of a dog—development of domesticity, faithfulness, love, morality, and all the amenities and virtues." London's view of evolution here is distinctly different from the classic Darwinian account, for he goes on to remark that "I am an evolutionist, therefore a broad optimist . . . my love for the human (in the slime though he be) comes from my knowing him as he is and seeing the divine possibilities ahead of him. That's the whole motive of my 'White Fang' " (1990, xv). Darwin's disavowal of a direction or teleology for evolutionary processes had disabled any such possibility of optimism in his own works. London's theories, however, permit of astonishing extremes of development. Initially White Fang has an "outlook [that] was bleak and materialistic. The world as he saw it was a fierce and brutal world, a world without warmth, a world in which caresses and affection and the bright sweetness of the spirit did not exist" (192). However, it was London's creed that "Every atom of organic life is plastic. . . . Let the pressure be one way and we have atavism . . . the other the domestication, civilization. I have always been impressed with the awful plasticity of life and I feel that I can never lay enough stress upon the marvelous power and influence of environment" (xv), and that "every atom" applies to mental and spiritual as much as to physical makeup. Thus White Fang develops not only physically, becoming "quicker of movement than the other dogs, swifter of foot, craftier, deadlier, more lithe, more lean with iron-like muscle and sinew, more enduring, more cruel, more ferocious, and more intelligent" (182), but also mentally, learning to appreciate that although Weedon Scott may be physically weaker than the roughnecks, they recognize his social superiority and keep their hands off him. In White Fang's perceptions, "As the days went by, the evolution of *like* into *love* was accelerated" (249).

In this view of evolution as a potentially two-way process and the concomitant lack of emphasis on fixed class stratification, London differs very notably from his British contemporaries. Conan Doyle's *Hound of the Baskervilles*, for instance, also centers on a dog, albeit to very different effect, and also registers a considerable interest in evolution and heredity, as might indeed be expected from its medically-trained author. But Conan Doyle differs markedly from London in his insistence on the narrative as illustrating the working-out of a divine plan (Conan Doyle 1996, 18). This difference of emphasis is the primary result of comparing London with Conan Doyle's friend and contemporary, Rider Haggard. Very like Haggard in his own great personal interest in agriculture, London differed from him profoundly in his belief in the possibility of change and improvement, whereas the fatalistic Haggard stressed continuity.[1] Moreover, the two men are also sharply differentiated in their attitudes about the question of an overall purpose in the universe. Haggard certainly makes copious use in his writings of the trappings of evolutionary theory. However, as the concluding lines of the poem that appears on the title page of *Allan and the Ice-Gods* (1971) reminds us, there may be more than one explanation of events, and we cannot be sure which is the true one. "Some call it Evolution, / And others call it God," says Carruth. The novel itself opens with a similar debate by Allan Quatermain before he meets a palaeolithic version of himself. This tribe worships a frozen mammoth and lives by "the doctrine of the survival of the fittest and the rights of the strong over the weak, as Nature preaches them in all her workings" (Haggard 1971, 74).

Indeed both *Allan and the Ice-Gods* and *The Ancient Allan* (1920) look almost like a direct riposte to London. In the former, Allan Quatermain reexperiences his past as paleolithic man; in the second, he recalls himself as "half human," as "black man," and as Egyptian of Ethiopian descent. Both thus come very close to *Before Adam* (1907), and *Allan and the Ice-Gods* in particular shares with *Before Adam* descriptions of the advance of ice to a place where it had not previously been seen and a sense of the resulting molding of the landscape by glaciation. Furthermore, linking these two works are accounts of attack by seaborne invaders, the hero's challenging of a huge and dominant leader who monopolizes the supply of women, and crucial plot developments motivated by the love of a woman from outside the tribe.

There are, however, striking and very significant differences. Haggard, whose already nascent interest in spiritualism was much intensified after the early death of his beloved only son, constantly insists both on the workings of a divine plan and on the persistence of individual human identities across gaps of time and after death. Though Allan Quatermain may have had another name in his earlier incarnations, he is always recognizably himself, and finds that his destiny is always inextricably interwoven with the same small, tight-knit group of people. London's protagonist, in contrast, is identified only by a soubriquet in his avatar and has no name or other identifying features in his modern descendant. (Indeed the possibility that the novel is autobiographical is left titillatingly open, and critics of London have not been slow to point to the parallels between London's own illegitimacy and Big Tooth's ignorance of the eventual fate of his father. Moreover, the elusiveness of the Swift One certainly has something in common with the alleged behavior of London's second wife, Charmian Kittredge). Yet, apart from dreams, the novel seems to contest the idea of continuity between the modern world and the primeval. The only point of contact between his otherwise entirely severed and discontinuous lives is a resemblance between Marrow-Bone and his father's gardener, which seems primarily attributable to the fact that both are old.

Haggard is also invariably at pains to stress that the ancient Allan was just as inventive and ingenious as the modern one (and just as fine a shot, even though his weapons were different). The narrator of *Before Adam*, by contrast, records that Big Tooth and the Folk "were without weapons, without fire, and in the raw beginnings of speech. The device of writing lay so far in the future that I am appalled when I think of it" (8: 2);[2] similarly, "Our evolution into cooking animals lay in the tight-rolled scroll of the future" (8: 5). The narrator repeatedly stresses that they show no initiative or constancy of purpose, and their only discoveries and advances are made entirely by accident—indeed London remarked that he wrote the book specifically to show "that in a single generation the only device primitive man, in my story, invented, was the carrying of water and berries in gourds" (Labor 1974, 106). Not without reason does London's narrator observe, "As I look back I see clearly how our lives and destinies are shaped by the merest chance" (*Before Adam,* 12: 3). Although Buck in *The Call of the Wild* has race memories of times with a caveman, "shorter of leg and longer of arm, with muscles that were stringy and knotty rather than rounded and swelling" (41), Big Tooth and the Folk have not yet reached even that stage of incipient civilization. There are no dogs in this novel except for the wild ones that chase Big Tooth up the cliff and almost eat him (12: 1) and the puppy that Big Tooth attempts to tame but which Lop-Ear eats, an episode that the narrator uses to stress yet again the workings of chance rather than design: "[t]o show you how fortuitous was development in those days let me state that had it not been for the gluttony of Lop-Ear I might have brought about the domestication of the dog" (8: 3). London's early men learn nothing so fast as do either Buck or White Fang.

The most notable difference from Haggard, however, is that London specifically denies (in *Before Adam*, at any rate)[3] that the phenomenon he is describing is reincarnation:

> I do believe that it is the possession of this other-personality—but not so strong a one as mine—that has in some few others given rise to belief in personal reincarnation experiences. It is very plausible to such people, a most convincing hypothesis. When they have visions of scenes they have never seen in the flesh, memories of acts and events dating back in time, the simplest explanation is that they have lived before. . . . But they are wrong. It is not reincarnation. I have visions of myself roaming through the forests of the Younger World; and yet it is not myself that I see but one that is only remotely a part of me, as my father and grandfather are parts of me less remote. (2: 2)

Haggard unequivocally accepts the idea of reincarnation, and even Conan Doyle flirts with it, making Holmes say in *The Hound of the Baskervilles* that "A study of family portraits is enough to convert a man to the doctrine of reincarnation" (145). London's narrative, however, even questions and problematizes the very possibility of the relationship between the narrator and his "other-self," as he calls Big Tooth.

One of the most remarkable features of *Before Adam* is the way in which it repeatedly insists on the lack of success of its protagonist's people: "We were the first of the Folk to set foot on the north bank of the river, and, for that matter, I believe the last.

That they would have done so in the time to come is undoubted; but the migration of the Fire People, and the consequent migration of the survivors of the Folk, set back our evolution for centuries" (11: 4). Like discourses of degeneration, London stresses the dark side of evolution; his narrative consistently frustrates any sense of progression and dwells instead on failure. The Folk as we know them during the course of the story are, ultimately, completely wiped out. Red-Eye survives, but reverts to the Tree People, to whom he had always seemed more properly to belong. Big Tooth himself survives, but he had not originally been part of the Folk. We are also explicitly told that the characteristics of his parents seem not in any sense to have been passed down: his mother "was like a large orangutan, . . . or like a chimpanzee, and yet, in sharp and definite ways, quite different" (3: 2), which does not sound even like Bigfoot himself, while his father "seemed half man, and half ape, and yet not ape, and not yet man. . . . There is nothing like him to-day on the earth, under the earth, nor in the earth" (3: 3).

Big Tooth's modern descendant wonders about the chain of descent:

> I, the modern, am incontestably a man; yet I, Big-Tooth, the primitive, am not a man. Somewhere, and by straight line of descent, these two parties to my dual personality were connected. Were the Folk, before their destruction, in the process of becoming men? And did I and mine carry through this process? On the other hand, may not some descendant of mine have gone in to the Fire People and become one of them? I do not know. There is no way of learning. (18: 3)

Darwin's theory of descent with modification has raised such issues but has not provided absolute answers to these questions.

As always in the writing of London the lifelong socialist, however, it is environment rather than heredity that proves to be the really crucial factor. It is not *any* child of Big Tooth's that could transmit his memories, but only one born in particular circumstances. Early in the narrative we are assured—on the authority of a college professor—that the "racial memory" of the falling dream arose because "a terrible fall . . . was productive of shock. Such shock was productive of molecular changes in the cerebral cells. These molecular changes were transmitted to the cerebral cells of progeny, became, in short, racial memories" (2: 1). This is, essentially, straightforward Lamarckism: a child born after such an event will inherit the altered metabolism and genetic makeup of the affected parent, while one born before will not.

This seems to be the explanation for the narrator's otherwise apparently puzzling assumption that he himself must be descended from a putative child born *after* the migration rather than the definite one to whom we have already been introduced, who had been born before it: "[t]he Swift One and I managed to bring up one child, a boy—at least we managed to bring him along for several years. But I am quite confident he could never have survived that terrible climate" (18: 2). The phrasing here is evasive: the child would, we are told, certainly have died if they had stayed, and even though they didn't, we might nevertheless be led to assume that his lifespan was limited to the "several years" mentioned. In any case, no further descendance is envisaged for him, for the narrator explicitly says of the cave where they settled after their final migration, "[h]ere the Swift One and I lived and reared our family. . . . And here must have been born the child that inherited the stuff of my dreams, that had moulded into its being all the impressions of my life—or of the life of Big-Tooth, rather, who is my other-self, and not my real self" (18: 3).

In short, even though the narrator cannot actually know that there ever was a later child, he prefers to believe in one rather than to imagine his earlier-born son, whom he has already in effect written off, as a link in the chain of his own later ancestry. In part this may be, as I suggested earlier, traceable back to Lamarckian ideas about the possibiliy of the transmission of acquired characteristics to offspring—only a child born after the migration could have transmitted to its own subsequent offspring any memories of that migration. But there seem also to be other forces at work. Throughout the narrative, the narrator has consistently pointed up the primitivism and lack of initiative of the Folk with whom Big Tooth allies himself and of whom, we suppose (though we can never be quite sure of this), he is ethnically a part. In one way, this can be fairly obviously related to the desire to produce an effect of deliberate contrast with the savage racism that so often characterized British imperialist co-options of evolutionary theory. For Haggard, the Zulus, though heroic, are developmentally on a par with the ancient Greeks; they are living anachronisms who must adapt or die—with their ability to do the former in considerable doubt since in 1864 the British Anthropological Society had declared that black children do not develop beyond the age of twelve. As Lyn Pykett remarks, "study of 'primitive' cultures proved extremely useful to the European domination of the 'dark races' in the Age of Empire" (1995, 27). Here, though, the already ideologically charged term "Folk" evokes no sense of patriotism or exclusivity, but indeed becomes virtually synonymous with idiocy: "the Folk in that day had a vocabulary of thirty or forty sounds" (4: 1), and "[w]e were ever short-sighted, we Folk" (11: 1).

Indeed it seems ultimately that Big Tooth's own progeny may survive essentially because it is also the child of the Swift One, who, he thinks, "may have been related to the Fire People" (10: 3). For his own kind, he is not only not triumphalist, he envisages no future at all. Unlike the Haggardian sense that those who are masters have always been masters and have a destiny to civilize, London's Folk are the accidental victims of colonization rather than its instigators, perhaps reflecting American resentment at their own past history as a colony. Moreover, the suggestive similarities between the Folk and Native Americans[4] touch on sensitive territory (there are, for instance, similar practices of nomenclature, and a similar pattern of suffering at the hands of technologically more advanced incomers). This comparison might well have led London to wish to present the differences between peoples in terms less confident than the imperialist adventurer Haggard, who characterized them as innate and ineradicable differences in capacity and destiny. It is true that White Fang thinks white men are a superior race to Indians (1990, 209) and that Alfred Kazin, not without reason, calls London himself "a prototype of the violence-worshiping Fascist intellectual if ever there was one in America" (1942, 111); however, Stoddard Martin argues that while "London belongs to that school which, following the theories of Darwin and Spencer, believed in an evolutionary hierarchy of races," he "was by no means so racially motivated that he would always portray his Anglo-Saxons as heroes and others as inferior villains. . . . Numerous examples from the Klondike and Hawaiian tales show London casting Indians or islanders in sympathetic roles" (1983, 41 and 46).

I think there is also another and more important reason for the mystification of the line of descent. For London, identity never exists in isolation but, as he often averred, is shaped by environment as much as heredity. Both Buck and White Fang change only because of those whom they encounter; had Buck never been stolen, or White Fang never met Weedon Scott, their behavior would have remained unaltered. In a crucial sense, then, Big Tooth's child is also the collective offspring of all the various individuals and tribes who have shaped Big Tooth's sense of his own identity. To the extent that environment thus *is* heredity, individual heredity is unimportant. Big Tooth's son is the child of the time, just as whiteness, in London's imaginary, is ultimately comprehensible only in terms of redness.

And this in turn takes us back to what seems to me to be the ultimate question about London's evolutionary fiction, the extent to which it is not only all men who influence an individual man's identity, but all animals. Far more subversive than *Before Adam*'s differentiation from Haggard—which would in any case be only retrospectively apparent—is its difference from a more prestigious account of man's earlier history, the Bible. The novel's very title, with its subversive suggestion of a quasi human identity before the creation of Adam, entirely rules out the possibility of an accurate Biblical account of the origins of humanity. It is true that there are, as in so much British writing influenced by theories of evolution, occasional echoes of Milton who, along with Darwin, had been one of the two authors who had formed London's staple reading in the Klondike. For example, Big Tooth's mother, like Eve, "wore no clothes—only her natural hair" (3: 2) and he himself, like Adam, longs for the companionship of another human (2: 1), which is eventually granted in his marriage with the Swift One. But his account of his own origins is resolutely materialistic: "some strains of germplasm carry an excessive freightage of memories—are, to be scientific, more atavistic than other strains; and such a strain is mine" (2: 2). And the epigraph to chapter 1 is equally uncompromising in its insistence not only on simian but, ultimately, on marine ancestry: "These are our ancestors, and their history is our history. Remember that as surely as we one day swung down out of the trees and walked upright, just as surely, on a far earlier day, did we crawl up out of the sea and achieve our first adventure on land."

Here is what might well at first appear the darkest and most dangerous area of London's text, the implication that made his dog stories so much safer ideological territory, like Darwin's decision to describe the breeding of pigeons rather than people in *The Origin of Species*. The idea of the chain of descent touches not only on the question of human origins, and the extent of their continuity with other animal life, but the related and far more frightening one of the eventual direction of the human race. In *Before Adam*, the question is implicitly answered by the bitter irony of the narrator, the only nugatory remnant of the Judaeo-Christian heritage, with his propensity to inexplicable and apparently unwarranted guilt, which prevents him from revealing his dreams to his parents: "I was afraid to tell. I do not know why, except that I had a feeling of guilt, though I knew no better of what I was guilty" (3: 5). Ultimately, then, even if man has progressed, it may only have been in the direction of a problematic and unnecessary complication that has belittled and bedeviled as much as it has enriched him. This, though, is where London's theories of race and speciation offer hope. If a man has listened to the voices within him and still has access to both civilized and savage modes of being, he can, like Buck in *The Call of the Wild* and the narrator in his dreams, always change back again. This is possible especially if, like Big Tooth and the heroes of the Northland stories, he cements his new-chosen identity by miscegenation. For London, who liked his own wife to call him Wolf, the man who remains in touch with both redness and whiteness will always retain something of the dual potential of the animal that can choose whether to be dog or wolf.

Notes

1. On the relation between London and Haggard, see also Gair 1997, 53.
2. Quotations from *Before Adam* (2000) are from the version available on the internet at http://sunsite.berkeley.edu/London/Writings/BeforeAdam/chapter1 .html. Each chapter has an appropriate variation of suffix, which provides a separate link for each of the eighteen chapters. In the interest of clarity, I therefore include only chapter and page numbers in citations within the text.
3. He had, it seems, changed his mind on this by the time he published *The Star Rover* in 1915.
4. For comment on this similarity and on London's sympathy with the Folk, see for instance Crow 1996, 50.

References

Auerbach, Jonathan. 1996. *Male Call: Becoming Jack London*. Durham, N.C.: Duke University Press.

Conan Doyle, Arthur. 1996. *The Hound of the Baskervilles*. Harmondsworth, Eng.: Penguin.

Crow, Charles L. 1996. "Ishi and Jack London's Primitives." In *Rereading Jack London*, edited by Leonard Cassuto and Jeanne Campbell Reesman. Stanford, Calif.: Stanford University Press.

Foner, Philip S. 1947. *Jack London: American Rebel*. New York: The Citadel Press.

Gair, Christopher. 1997. *Complicity and Resistance in Jack London's Novels*. Lampeter: The Edwin Mellen Press.

Geismar, Maxwell. 1954. *Rebels and Ancestors: The American Novel, 1890–1915*. London: W. H. Allen.

Haggard, H. Rider. [1927] 1971. *Allan and the Ice-Gods*. Reprint, London: Hutchinson.

Kazin, Alfred. 1942. *On Native Grounds: An Interpretation of Modern American Prose Literature*. New York: Harcourt Brace.

Labor, Earle. 1974. *Jack London*. New York: Twayne.

London, Jack. 1990. *The Call of the Wild, White Fang and Other Stories*. Edited by Earle Labor and Robert C. Leitz, III. Oxford: Oxford University Press.

———. *Before Adam*. http://sunsite.berkeley.edu/London/Writings/Before Adam. Accessed 15 June 2000.

Martin, Stoddard. 1983. *California Writers*. Basingstoke, Eng.: Macmillan.

Pykett, Lyn. 1995. *Engendering Fictions: The English Novel in the Early Twentieth Century*. London: Edward Arnold.

Evolving into Violence: Poor White Humor in T. S. Stribling's *Teeftallow*

Debra Beilke

Introduction

T. S. STRIBLING'S 1926 NOVEL *TEEFTALLOW* PORTRAYS, AMONG OTHER "amusing" antics of the Tennessee hill folk, the lynching of one man, the beating to near death of the protagonist Abner Teeftallow, and the expulsion from town of Abner's girlfriend for engaging in premarital sex, all at the hands of the local "white caps." Furthermore, the novel ends with the unpunished murder of the local hero-trickster, Railroad Jones, after he legally cheats the entire population of their savings, his best friend out of his job and good name, and the protagonist Teeftallow of his rightful inheritance. Despite these and other horrors, however, the tone of this novel is one of detached comedy. T. S. Stribling's uneasy humor, combined with his evolutionary rhetoric, suggests that these southern hillbillies are so peculiar, so biologically distinct from the implied readers, that we can laugh at their violent antics rather than be deeply moved by the harm they cause.

Although almost unread today, Tennessee native T. S. Stribling was a popular, prolific, and critically acclaimed novelist in the 1920s and 1930s. *Teeftallow* was, for example, a main selection of the Book-of-the-Month Club in 1926, and Stribling won the Pulitzer Prize for his 1932 novel *The Store*. Writing mostly in the vein of critical realism, Stribling focused in his fiction on the social issues of the South, such as the narrow-mindedness and violence of small-town Tennessee culture depicted in *Teeftallow*. His penchant for social critique has led some critics—both admirers and detractors—to place him in the liberal, reformist camp of 1920s American literature.[1] A closer examination of the novel's humor and evolutionary rhetoric, however, suggests that the author's political views are conservative rather than reformist. More specifically, *Teeftallow* discursively saddles the responsibility for southern violence on the sociobiological evolution of the southern "white trash." It is not surprising that Tennessee poor whites do not understand the theory of evolution, Stribling's satire implies; their evolutionary lag has resulted in their inherent feeblemindedness. Furthermore, his novel suggests that attempts by progressives to remedy the injustices of the South are pointless because the poverty and violence of the poor whites are a natural result of heredity and are thus immune to externally-imposed reform.

Attitudes towards Poverty in the South

Published in 1926, *Teeftallow's* evolutionary comedy was particularly well-received because of its reverberations with the infamous Scopes "monkey" trial of 1925. This trial gave the South a derisively comic reputation for backwardness and religious fanaticism. Taking advantage of the cultural shorthand that associates anti-evolutionary thinking with southern backwardness, Stribling places the theory of evolution at the center of the themes and comic technique of this novel. Publicity over the trial created a receptive arena for his theory that southern backwardness, especially its violence, is caused by the evolutionary development (or lack thereof) of the southern poor whites.

While Stribling's particular comic rhetoric is somewhat unusual, he is not the first author, of course, to evoke laughter at the expense of southern poor whites. In fact, the gruesome humor of his novel fits within a long tradition in southern letters; this tradition, as Sylvia Jenkins Cook observes, associates poor whites with "humor, gothic horror, and 'such poverty that it was a matter of jest'" (1976, 91). But while southerners love to laugh at poor whites, this group's definition is notoriously hazy. Economic indicators are inadequate; while virtually all "poor whites" are materially deprived, not all whites who are poor are "poor whites." Rather, the label connotes a moral and sometimes biological inadequacy, and it is usually associated with laziness, degeneracy, immorality, and extreme ignorance.

While Stribling's representation of poor whites has much in common with representations of this group elsewhere in the South, in *Teeftallow* he focuses on one subdivision of the southern poor white category: the Appalachian "hillbillies," a group with its own history of representation. In *Southern Folk, Plain and Fancy*, sociologist John Shelton Reed identifies hillbillies as "the last acceptable ethnic fools," a group that has "amused non-Southerners and upper-class Southerners alike since the days of Robert Beverley and William Byrd, going on three hundred years ago" (1986, 43). Although, according to popular representations, the hill folk are similar to their lowland poor white counterparts in their brutal poverty, ignorance, and aversion to work, they differ in their fierce independence, their loyalty to their mountain homes, and most important for my purposes, their tendency towards lawlessness and violence. In *Appalachia on Our Mind*, Henry Shapiro argues that "the tendency of mountaineers to engage in feuds, and more specifically the practice of private justice through ambush or 'bushwhacking' . . . had already become a part of the mythology of Appalachian otherness by 1900" (1978, 104).

Many commentators on Appalachian otherness, influenced by social Darwinism and popular genetic theory, attributed the "peculiarity" of the Appalachian dwellers to their biology. They were, according to this theory, an inferior ethnic group who descended from the "degraded stock, paupers and criminals, the 'scum of London streets' transported to colonial Virginia" (Shapiro 1978, 95) and who eventually drifted westward into the mountains where their peculiar characteristics made them more fit to live. Stribling mines this literary tradition of southern poor white hill folk as a lawless, violent, and comically inferior group of Americans, a group that he portrays as "naturally" violent because of a long process of institutional and biological evolution.

Although Stribling's poor white satire participates in a long tradition, the specifics of his rhetoric are informed by and contribute to debates specific to the historical context of the mid-1920s, the period in which *Teeftallow* (1926) was written and published. Not only did the 1920s in the United States give rise to debates over the teaching of evolution, but this decade also witnessed a rise in "one hundred percent American" nativist sentiments as well as a national focus on the South's violent lawlessness. Nativist sentiments in the United States crested in the 1920s. While the ideology of WASP supremacy in America did not originate in the 1920s, the dramatically increased immigration from Eastern and Southern Europe as well as Asia in the late-nineteenth and early-twentieth centuries incited a reactionary movement touting the putative superiority of White Anglo-Saxon Protestants. The South was no exception to this trend. Indeed, southern boosters often pointed with pride to white southerners' presumed ethnic homogeneity, seizing upon the national temper to promote the superiority of the southern "stock." Historian George Tindall summarizes the logic behind this ideology as follows: "If the test of Americanism was native birth of Anglo-Saxon ancestry (or Scotch-Irish—either sufficed), it followed that the South was the 'most American' of all regions, derived from the 'best stock'" (1967, 185). Many white southerners prided themselves on their ethnic homogeneity; unfortunately this source of pride facilitated racism and xenophobia, which led to a number of disturbing social trends that inform Stribling's work.

One infamous reaction to the changing face of the nation was the reincarnation of the Ku Klux Klan (a version of which appears in *Teeftallow* as the "whitecaps"), which terrorized immigrants, Jews, Negroes, and Catholics. At its zenith in 1925, the KKK, according to some estimates, boasted as many as eight million members nationwide. While the Klan spread across the nation, its influence, according to Tindall, "was nowhere so pervasive or violent as the South. Its progress across the region left a trail of threats, brandings, floggings, emasculation, and murder" (1967, 192). Concurrent with the revival of the KKK was the continuing southern tradition of lynching, the victims of which were overwhelmingly (ninety percent) African-American. While lynching was not exclusively a southern phenomenon, this type of murder did occur primarily in the South.[2] Lynching, of course, began much earlier than the 1920s, having been pervasive in the South since the late nineteenth century. However, the mounting efforts against lynching begun after 1910 by organizations such as the NAACP, and the Commission on Interracial Cooperation brought national attention to this southern blight.

All the national publicity on the Klan, on lynching, and on other southern problems crystalized the association in the national psyche between the South and lawlessness. The benighted South gained prominence in the national press, which ground out "one Southern abomination after another" (Tindall 1967, 212). Donald Davidson aptly summarizes the national image of the South during the 1920s when he writes, "The South—so the tale runs—is a region full of little else but lynchings, shootings, chain-gangs, poor whites, Ku Kluxers, hookworm, pellagra, and a few decayed patricians whose chief intent is to deprive the uncontaminated, spiritual-singing Negro of his life and liberty" (quoted in Tindall 1967, 215).

This national image of a lawless and degenerate South led many southerners, not surprisingly, to a defensive stance and with it a concomitant desire to exculpate themselves from responsibility by finding a scapegoat. While white southerners often chose African-Americans as scapegoats, it was difficult to blame them for KKK violence and other forms of lynching. Next in line for blame stood the poor whites. This group was frequently blamed for many pervasive southern problems, including its lawlessness. In 1902, for example, southern classics professor Andrew Sledd published an anti-lynching article in *The Atlantic Monthly,* which asserted that "our lynchings are the work of our lower and lowest classes" (Williamson 1984, 260). And in *The Tragedy of Lynching,* Arthur Raper notes that known active lynchers are generally unemployed and without property, and that "in the rural communities the more shiftless types of white farm tenants and wage hands were most in evidence" (1933, 11).

Historian Joel Williamson calls this tendency to find the root of all southern evil—particularly extreme racism and violent lawlessness—in the ignorant white masses the "grit thesis" (1984, 292). Debunking the "grit thesis," Williamson demonstrates that while the dispossessed do tend to express their prejudices and frustrations in overt physical violence, upper-class prejudice "is often manifested in more subtle forms of economic, social, psychological, educational, and judicial manipulation," which is as violent as "guns, whips and bombs" (294). Lynching and other forms of extralegal violence generally involve communities rather than individuals; passive onlookers and the system of justice that refuses to prosecute the murderers share partial responsibility for the murders. Given such a system, blaming the "lower orders" of society for the violence is obviously a defensive mechanism, a means of finding a scapegoat with little political power to fight back.

But while the poor whites' lack of economic and political power renders them easy targets for vilification, the fact that they are white poses a problem for those who believe in a genetic basis for human behavior. How can the poor whites be so awful, so genetically predisposed to violence, poverty, laziness, and other unsavory characteristics, if they come from the same "superior" racial and ethnic heritage as ruling-class WASPs? This question puzzled a number of social commentators. The preponderance of southern "po' white trash," with their bizarre and degenerate ways, embarrassed proponents of "superior" Anglo-Saxon racial purity. Author and publisher Walter Hines Page wrote in 1912, for example, that "the Southern white people are of almost pure English stock, [and thus] it has been hard to explain their backwardness, for they are descended from capable ancestors and inhabit a rich land" (quoted in Williamson 1984, 451). And in *The Passing of the Great Race*, prominent eugenicist Madison Grant puzzled over the preponderance of "white trash" found among the Nordic stock of Kentucky and Tennessee:

> The poor whites of the Cumberland Mountains in Kentucky and Tennessee present a more difficult problem, because here the altitude, even though moderate, should modify the effects of latitude and the climate of these mountains cannot be particularly unfavorable to men of Nordic breed. There are probably other hereditary forces at work there as yet little understood. (1921, 39)

Many writers have "solved" this conundrum of degenerate white stock by representing "white trash" as a separate genetic category from "good" WASPs, essentially positing a biological basis for class distinctions. W. J. Cash criticizes this way of thinking in his classic *The Mind of the South.* He attempts in this book to explode the myth that he found still prevalent during his time: the belief in a biological distinction between the master classes (planters and their descendants) and the materially deprived poor whites, whose shiftlessness and criminality were believed to be "inherent in the germ plasm" (1941, 6). According to this myth, as Cash relates it, the Cavaliers of the ruling class are "naturally" fit to rule because they are "descended from the old gentlefolk who for many centuries had made up the ruling classes of Europe" (ix). The myth would have it that the "white-trash," on the other hand, are physically inferior to the ruling class, "having sprung for the most part from the convict servants, redemptioners, and debtors of old Virginia and Georgia, with a sprinkling of the most unsuccessful sort of European peasants and farm laborers and the dregs of the European town slums" (ix).

T. S. Stribling's rhetoric also implies a biological basis for poor white "otherness." Stribling's main goal in this novel, however, is to satirize, to evoke laughter from the antics of the mountaineers. A close examination of Stribling's humor reveals a deep uneasiness towards the poor whites he lampoons. It is easier to laugh at others' misfortunes if one feels distant from and superior to them. Knowing this, Stribling strives to increase the psychological distance between his characters and his implied readers by positing the biological inferiority of the Tennessee hillbillies he describes. Stribling's comic strategy invokes readers' laughter while simultaneously setting up poor white hillbillies as scapegoats to explain away the violence infecting the South. And yet, the markers of the putative biological inferiority of poor whites are so slippery, so subjective, that there remains a significant amount of horror attached to poor white humor. Readers can more easily laugh at these characters if they seem distant, biologically unlike "us." When that distance diminishes and readers see the similarities between poor whites and themselves, their antics tend to evoke horror rather than humor.

TEEFTALLOW

The plot of *Teeftallow* focuses on approximately two years in the life of Abner Teeftallow, an orphan who was raised by the overseer of the county poor farm, after he is "thrown out upon the world to make his own living" (1926, 28) at age seventeen. Abner gets a job constructing a new railroad, befriends several village men, impregnates his girlfriend Nessie, and generally learns the ways of village life. But although the plot focuses on Abner's coming-of-age, this satiric novel is less a bildungsroman than a pseudo-anthropological study of the Tennessee hill folk. As Wilton Eckley correctly observes, "character development or motivation is not what [Stribling] was interested in when he wrote *Teeftallow*. He was more interested in examining hill-country mores with the eye of a reporter and in recording what he saw" (Eckley 1975, 53).

The main subject of this novel, then, is the hill folk as a group, rather than any particular individual. Scattered throughout the text are the narrator's observations on traits of *the* hill folk (of which Abner is but one specimen), eliding individual distinctions in favor of one monolithic type. We learn, for example, in keeping with poor-white stereotypes, that the "goal of the hill people [is] a state of complete idleness" (Stribling 1926, 287). Other inborn traits are the "hill instinct to trade . . . it's automatic with them" (52), "the stoic indifference of all hillmen for all hill wives" (93), and "the hill-born instinct . . . to associate wickedness with everything pleasant, graceful, or beautiful" (291).

While Stribling pokes fun at a variety of Appalachian traits and customs, he particularly satirizes the hill system of justice, especially the locals' reliance on lynch law rather than the ludicrously ineffective official legal system. For example, one character justifies the town's reliance on lynch law by arguing that "This ain't a case fer the law; there's been a crime committed" (174). Although "justice" and "the law" are theoretically synonymous, in Abner's world they are in fact antithetical. As Abner muses, "the people here do this [extra-legal violence] . . . because the law won't do nothin' to nobody" (187). But while Stribling depicts a serious problem in this novel, his purpose is not, as some claim, to instigate social reform. Rather, his novel dictates political passivity: there is nothing we can do to reform society except to wait for biological change. In an interview, Stribling once summarized his views on change:

> Every society in every day of its existence is continually changing—and it is not changing according to any one man's conception. It isn't a logical change; it's a biological change. And that goes on, and the logical concept pursues this biological change and makes different futures that men think the biological change will pursue. But it never pursues the one that they think. (Eckley 1975, 110)

Stribling transmutes this philosophy, which posits biology as the primary instigator of social change, into fictional form in *Teeftallow*. Given his emphasis on the biological rather than cultural underpinnings of change, it is not surprising that evolution plays an integral role in his work.

The author establishes evolution as a thematic framework early in the novel, which primes the reader to interpret the characters through an evolutionary lens. A key scene in the third chapter, for example, resonates with issues central to the Scopes trial. While waiting at the courthouse for his foster-father's legal proceeding, Abner witnesses a hearing in which Brother Blackman, a local evangelist, gives a speech to the justices of the court. Blackman tries to persuade them to sign an anti-evolution petition by using revivalist rhetoric:

> He proceeded in his ponderous voice, and inquired slowly and solemnly, "My frien's, do you b'lieve your great gran'daddy was a monkey?"
>
> He paused, then with the revivalist's trick, shouted the same question with a different stress at the top of his lungs, "Do you *b'lieve* your great gran'daddy was a monkey?"
>
> This jarred the nerves of his audience. The preacher brought down his fist on the chancel rail with a sounding blow, "Is there a *man* in the sound of my voice that *b'lieves* his great gran'daddy was a monkey? . . . Oh, brethren, don't you *know* the Bible says man was

> made in the image of *God*! Then how *can* he be made in the image of a monkey?" . . .
>
> The justices looked at [the petition] rather blankly and signed one after another. One of the court hesitated a moment. "Professor Overall," he asked, "does our present school books teach there ain't no God an' our gran'fathers was monkeys?"
>
> Professor Overall rolled his prominent eyes on the questioner reprovingly. "They certainly do, Brother Boggus. You can take my word as a teacher and a scholar."
>
> "I jest wanted to know," said the justice in a chastened tone, and signed his name hurriedly. (1926, 26)

This passage exemplifies the way *Teeftallow*'s humor hinges on the theme of evolution: we laugh at these hill folk because they are so ignorant that they misunderstand Darwinian thought to the extent that they accept Blackman's skewed interpretation of it. Stribling's mockery of anti-evolutionist Tennesseans was nothing unusual, though; in many ways his comic derision echoes the national coverage of the Scopes trial. Stribling is somewhat different, though, in his use of the theory of evolution to suggest that poor white hillbillies are not only *culturally* backward compared to the implied middle-class reader, they are also *biologically* inferior.

Stribling hints at an evolutionary lag in the hill folk by emphasizing the visual markers of their backwardness. Describing a group of hillmen, for example, the narrator informs readers that "the whole group had the gnarled, almost grotesque faces developed by generations of illiteracy" (279). And we learn from Adelaide, Abner's rich, second girlfriend, that Abner exudes a sort of virile, prehistoric appeal; she tells him that "it's nice to talk to such a cave-looking man after the usual dancing partner" (294). But although Adelaide finds Abner physically attractive, he has not yet evolved enough to dance publicly with her: " 'Oh, no, dear, you won't do to dance in polite society for several generations yet to come. Now, your great-great-grandson may be able to dance with a lady in a perfectly gentlemanly manner, like Buck' " (299). Abner's evolutionary lag is typical of the hill folk. There was, for example "something *primitive* in the way [Tug Beavers] halted his chewing and swallowing to stare through the glow of the lamp at the girl beyond" (40; emphasis added).

Stribling's condescending attitude towards his fictional subjects suggests that their backwardness is not merely a matter of ignorance, which can be remedied by education and other forms of "uplift." Rather, *Teeftallow* suggests that education would be fruitless because, like animals, the hill folk are governed completely by blind instinct and are incapable of logical reasoning or insight into the source of their problems. For example, Abner becomes violently frustrated in his work on the railroad, but, as the narrator informs us, "Why he should work so fiercely, why objurgate so filthily and indecently, he had not the faintest idea. All the teamsters did the same thing *and none of them had the slightest notion of why they were so stirred and so bitter*" (45; emphasis added). The hill folks stubbornly resist "edjercation" because, according to local wisdom, literacy makes "a complete fool out of [people]" (7). Indeed, the most powerful man in town, Railroad Jones, is illiterate. But he succeeds because he possesses vital skills for his environment—devious cunning and the ability to out-trick his opponents.

The hill folks' instinct towards trickery and deviousness has evolved because of the specific geographical and historical conditions of their ancestors, the frontiersmen/Indian fighters. The narrator frequently refers to the early pioneers and their attacks on the Indians. He, for example, attributes a particularly skewed logic in hill law to the influence of the Indian fighters: "Such a fantastic twist to hill morals was a relic of the old Indian fighters who first settled the country" (170). Furthermore, when Railroad Jones tricks Ditmas out of a large territory of timber, Abner laughs appreciatively because "[i]t was an exquisite game, this slipping up on the blind side of a purchaser and picking him off as the first pioneers had crept up on the Indians and picked them off" (280). And while observing the courtyard scene in a moment of leisure, an occupation that does not require any particular caution, Abner "stood as still almost as the trunk of the tree itself, for he was only a generation removed from Indian fighters and wild-game hunters, and the woodsman's manner of observing still clung to him" (10).

By comparing the hill folk to Indian fighters, Stribling suggests a genetic resemblance between the Appalachians and the American Indians. Although the hill folk are of Anglo-Saxon stock, battling the American Indians in frontier conditions forced them to adopt some Indian customs in order to survive. The textual evidence suggests that succeeding generations, in turn, inherited these environmentally-acquired characteristics. The characteristics acquired by his ancestors for environmental reasons, then, still biologically "cling" to Abner and the other villagers, suggesting a strong evolutionary basis for the customs of the Appalachians.

This particular evolutionary theory is, however, Stribling's own brainchild. In that environmentally-acquired characteristics are passed on genetically to succeeding generations, his theory echoes Lamarckianism, which, according to George Frederickson, is a view of human evolution that "permit[s] social influences to affect the genetic make-up of individuals and races" (1971, 314). Reformers cherished the theory of Lamarckianism (which had been scientifically discredited with the advent of Mendelian genetics by the turn of the century) because it implied that the benefits resulting from favorable environmental conditions would be passed on genetically to offspring. Stribling, however, rejects the reformist implications of this theory. Although his textual logic suggests that present traits have been inherited from a combination of environmental and biological traits, he illogically suggests that changes wrought by reformers will not be inherited by future generations. As such, he borrows from different evolutionary theories as they suit his needs, without any regard to their consistency.

Stribling's theory of biological change leads him to ridicule the idea of reform. The novel mocks well-intentioned but useless outsiders, usually northerners, who try to uplift the hill folk by imposing "edjercation," socialist reforms, labor unionism, and so forth on the hill folk. Stribling laughs at such reformers because they do not realize that change can only come slowly, through evolution, not swiftly through reform. His comic treatment of Shallburger, the northern socialist agitator, most clearly illustrates his rejection of social reform for poor white laborers. Shallburger has come to this area to organize the railroad construction workers into a union so that they can receive higher wages. His earnest efforts at uplifting the poor are greeted, however, with puzzled suspicion. Noting Shallburger's "queer sharp accent," Abner, "not understanding a word of what [he] said," surmises that "the fellow evidently was a Yankee—that is, a trickster" (1926, 71). Although Teeftallow and the other teamsters toy with the idea of labor agitation for awhile and even attempt to strike briefly, they finally reject unionism on the grounds of its stupidity: "The hill youth bristled at the recollection of Shallburger's stupid theories—they were so dead against common sense. He could not conceive how a sane man could think like that" (329–30). This quotation, however, suggests that it is the hill workers who lack common sense and sanity, since they are so dimwitted they cannot see what is in their best interests.

The comedy of Shallburger's continued efforts to raise the hill folks' class consciousness works both ways. On the one hand, his efforts provide more fodder for laughter at the expense of the hill folks' limited intelligence. After attending the organizer's meeting, for example, Abner explains to his friends that socialism is "a plan to divide up all the work into such little bits, everybody will be out of a job most of the time" (124). On the other hand, Shallburger's rhetoric and organizational attempts are comically ineffectual, especially compared to the masterful maneuverings of the illiterate Railroad Jones, whose "hill smarts" enable him to trick the entire population out of their life savings. The strike ends in defeat and nobody has any more class consciousness at the end of the novel than at the beginning.

The fact that nothing has fundamentally changed in this society—that social inequities and lawlessness persist as strongly as ever, despite the violent deaths of several people—is precisely Stribling's point. Any fundamental change will have to evolve slowly; there is nothing anyone can do to speed it up. Ditmas, the northern engineer, expresses Stribling's beliefs in a drunken monologue to which nobody pays any attention. After noting that contracts never mean what they say, Ditmas says he finally understands why lynch law prevails. The reasons for the dominance of the whitecaps, he argues, stem from a slow process of historical evolution:

> This-dis-this-disingenuous method of l-law and business here in South been a long gradual development, Abner—ver' long an' ver' gradual. I see it all before me, Abner—hist'ry of the South. . . . What result? Whitecaps, mobs, posses, lynchin's, burnin's, beatin's. . . . *But-but nobody's to blame. Since there's no law of right, there must be one of might.* Mobs and whitecaps, all over our nation. North and South, East and West, anywhere, ever'where—*but there's nothin' to do. . . . You're a citizen of the South, and of the United States, Abner, and don't you do nothin' a-tall about it, Abner—f' th' ain't nothin' to do.*" (285; emphasis added)

Nothing can be done about southern lawlessness because it is the result of slow, gradual development and is therefore immune to the type of sudden change that occurs with legal and social reforms.

The reference to slavery in this quotation raises an important question: since the focus of this novel is on southern mob violence and since in fact over ninety percent of lynchings were executed against African-Americans, why does Stribling make no mention of blacks in this novel? The erasure of race seems particularly striking since Stribling's first novel focused on racial injustice in the South. Perhaps he assumed that his readers would find white-on-white crime harder to identify with than white-on-black crime. Therefore, the "otherness" of the poor whites seems even more extreme, and the reader distances herself even further from the characters described. If, in turn, the readers attribute southern violence to the inherently degraded natures of the unruly white trash, then when white-on-black violence does occur it is easier to ignore the real reasons—the tensions arising from the racial caste system.

Far from advocating progressive reforms to eradicate poverty, ignorance, and violence, *Teeftallow*—with a political subtext revealed by its humor—promotes maintaining the social status quo. Southern violence, the novel suggests, is not so much a social problem as a biological problem: the genetic backwardness of Appalachian hillbillies. One implication of this rhetoric is to excuse the implied middle-class readers from any shared responsibility in the social problems depicted in this novel. In the end, however, the implied reader is left with a feeling of uneasiness. The novel is humorous only insofar as readers can distance themselves from the "poor white trash" being described. If we allow ourselves to laugh along with the author at the antics of these hillbillies, at an unconscious level we become complicit with ideology promoting a genetic basis for social and economic inequalities.

Notes

1. James Mellard, for example, categorizes Stribling as one of a group of socially committed writers "aiming for reform of historical conditions" (1985, 351); and Donald Davidson claims that Stribling propelled the reviewers of his books towards a "call to arms, that the South might be rescued from evil ways" (1934, 191).

2. As H. C. Brearley documents, "In the twelve southern states occurred nearly ninety per cent of the 1,886 lynchings that took place in the United States from 1900 through 1930" (1934, 679).

References

Brearley, H. C. 1934. "The Pattern of Violence." In *Culture in the South*. Edited by W. T. Couch. Chapel Hill: University of North Carolina Press.

Cash, W. J. 1941. *The Mind of the South*. New York: Vintage.

Cook, Sylvia Jenkins. 1976. *From Tobacco Road to Route 66: The Southern Poor White in Fiction.* Chapel Hill: University of North Carolina Press.

Davidson, Donald. 1934. "The Trend of Literature: A Partisan View." In *Culture in the South.* Edited by W. T. Couch. Chapel Hill: University of North Carolina Press.

Eckley, Wilton. 1975. *T. S. Stribling.* Boston: Twayne.

Fredrickson, George M. 1971. *The Black Image in the White Mind: The Debate on Afro-American Character and Destiny, 1817–1914.* Hanover, Conn.: Wesleyan University Press.

Grant, Madison. 1921. *The Passing of the Great Race or The Racial Basis of European History.* New York: Scribner.

Mellard, James. 1985. "The Fiction of Social Commitment." In *The History of Southern Literature,* edited by Louis Rubin, Jr., et al. Baton Rouge: Louisiana State University Press.

Raper, Arthur F. [1933] 1969. *The Tragedy of Lynching.* Reprint, New York: Arno.

Reed, John Shelton. 1986. *Southern Folk, Plain and Fancy: Native White Social Types.* Athens: University of Georgia Press.

Shapiro, Henry D. 1978. *Appalachia on Our Mind: The Southern Mountains and Mountaineers in the American Consciousness, 1870–1920.* Chapel Hill: University of North Carolina Press.

Stribling, T. S. 1926. *Teeftallow.* Garden City, N.Y.: Doubleday.

Tindall, George B. 1967. *The Emergence of the New South: 1913–1945.* Baton Rouge: Louisiana State University Press.

Williamson, Joel. 1984. *The Crucible of Race: Black-White Relations in the American South Since Emancipation.* New York: Oxford University Press.

The Origin of Story and the Survival of Character in Faulkner's *Absalom, Absalom!*

Alex Vernon

IN HER 1976 BOOK *DARWIN IN AMERICA: THE INTELLECTUAL RESPONSE 1865–1912,* Cynthia Russett observed that manifestations of American literary naturalism ranged "from the simple use of animal metaphors to the acceptance of a full-fledged philosophy of determinism" (174). A few decades after that movement's heyday, armed with historical perspective on his literary forbears, William Faulkner extended this range to what we might call the metanaturalism of his 1936 novel *Absalom, Absalom!*, which pushes beyond the conventions of naturalism by invoking a more thoroughly Darwinian evolutionary aesthetic.

Critical consideration of William Faulkner as an inheritor of naturalism is well established, though only a handful of scholars place Faulkner in an explicitly Darwinian context.[1] Of the 1925 Scopes "monkey" trial, Fred Hobson has written that "virtually every thoughtful Southerner had some response to the occurrences at Dayton" (1974, 148). Faulkner had more than just a passing interest in the debate. In December 1937, the year after the novel's publication, Faulkner "discussed Darwin with Mac, who had now begun reading in biology and anthropology under the spur of his stepfather's interest," and Faulkner gave him a copy of *The Origin of Species* for Christmas (Blotner 1974, 389, 396).[2] Even more than the original naturalist novelists, a number of whom had never read Darwin, Faulkner could talk the talk. In *Absalom, Absalom!* he hangs most of his Darwinian language on Thomas Sutpen and his progeny while more subtly incorporating the evolutionary model into the novel's very structure. The metanaturalism of this novel comments on the limits of naturalism as a philosophy through the narrative's more accurate evolutionary design and through the major (white) characters' inability to survive the text. This essay will also explore what the novel shares with post-Darwin Victorian culture; how it relates issues of memory, evolution, and adaptation; and how it might perform a kind of narrative miscegenation that unsettles the South's erroneous yet fundamental insistence on strict black-white speciation.

I

Sutpen builds his dynasty "plank by plank and brick by brick out of the swamp" (1990a, 28), working naked and covered in mud, the primal ooze, with "his band of niggers like beasts half tamed to walk upright by men" (4). References to his emergence from the swamp and mud pervade, such as Rosa's description of "that man who had struggled through a swamp with nothing to guide or drive him . . . and blundered at last without warning onto dry solid ground and sun and air" (134). His slaves are constantly made into beasts; one is stumbled upon in "the absolute mud like a sleeping alligator" (27). In this novel fictionally composed in 1910 and looking back into a Southern past on either side of the Civil War (and on either side of *The Origin of Species* [1859]), the historical perception of African Americans variously places them as animal rather than human or as a human race closer than European-Americans to primates in the evolutionary process. Yet while Sutpen's slaves, even with their spot of humanity, are perceived as "a good deal more deadly than any beast" (28), Faulkner indicates that Sutpen himself is more dangerous still, as his domination of them attests. As Sutpen rides in the carriage, his face appears "exactly like the negro's save for the teeth (this because of his beard, no doubt)," and wrestling with a slave, "both naked to the waist and gouging at one another's eyes as if their skins should not only have been the same color but should have been covered with fur too" (16, 21). Elsewhere their skin is the same color when they cover themselves in mud against the swamp's mosquitoes.

The beast-in-Sutpen motif comes to this novel straight from naturalism (see Cowley 1956 and Donaldson 1998). Over the course of a naturalist story, male characters often undergo a process of devolution, of regressing either to a beastlike self or a trapped, impotent self rendered pathetic by the overpowering forces of his social environment. Sutpen's story certainly matches this model. Rosa refers to him as "a certain segment of mud . . . in retrograde, reverse . . . descending" (139), and he dies with his plan in ruins, his grandson married to an "ape-like woman" dragged from some "two dimensional backwater" and "resembling something in a zoo" (166, 169). The last Sutpen, the great-grandson Jim Bond, is an idiot last seen—or rather heard—bellowing in the wilderness. It is Bond's progeny, according to Shreve, who will inherit the earth (302). Sutpen, with his initial successes in Haiti and Yoknapatawpha and on Civil War battlefields, appears to personify the pop Darwin maxim "survival of the fittest," yet his failure to survive demonstrates the speciousness of that phrase. Natural chance, not volitional control, determines survival. In literary naturalism, however, chance events couldn't occur because "all causes were knowable" (Mitchell 1998, 526). Sutpen's obsession with chance and its foiling of his scheme reveals both the limits of his will and the limits of naturalism—because for Darwin, evolution *is* chance. (The Darwin-invoked fear that there is no plan and the frantic search for an originating force explains Shreve's and Quentin's invention of the lawyer. They need human agency to explain things.)

Surprisingly, for all the symbolic value ascribed to Sutpen's mansion, the architect who designed it remains the novel's most critically neglected character. If the nameless slaves represent Sutpen's beastly half, the unnamed French architect represents his overly civilized soul, the Dr. Jekyll to the slaves' collective Hyde. They all appear in the wagon together, and it requires both architect and slaves to build Sutpen's house, Sutpen and his slaves "distinguishable from one another by his beard and eyes alone and only the architect resembling a human creature" (Faulkner 1990a, 28). Like the calculating Sutpen, the French architect possesses a "fatalistic and amazed determination. . . . with his air something between a casual and bitterly disinterested spectator and a condemned and conscientious ghost" (26, 28). It is no artistic coincidence that Sutpen tells his life story to General Compson while with his "beastly" slaves he chases, traps, and runs to ground his other symbolic self in the figure of the Frenchman. Immediately after Sutpen relates the youthful discovery of his methodical design to escape his family's plight, we return to the hunt to find that the architect "had used architecture, physics" to elude his hunters by swinging from tree to tree (193). Sutpen's narrative resumes with his acknowledgement of the original—albeit unintended—miscalculation, the first wife, and again we return to the hunt, where the architect has hurt his leg through "a mistake"—surely unintended—"in the calculation" (206). He has regressed into a "cave under the river bank" (206) with an indomitability, "a will to endure and a foreknowing of defeat but not beat yet by a damn sight" (207), which we readers see in Sutpen, especially postwar Sutpen with his children dead and vanished and him fighting time to produce an heir, not yet beat yet by a damn sight. In the end, the architect's science and Sutpen's method fail in their promise.[3] Faulkner may be making a literary joke as well, associating Sutpen with the father and architect of European literary naturalism, the Frenchman Emile Zola, whose aesthetic treated characters as explicable and predictable products of heredity and environment, as test subjects to be observed in the laboratory of narrative art.

The naturalist idea of a person's devolving into a beast, of a person's allowing the repressed animal self to emerge and conquer, follows from a widespread nineteenth century Darwin-derived correspondence between ontogeny and phylogeny, between the development of the individual and the development of the species. If we can trace the animal within the human, the logic goes, then we can find the animal in the individual. Faulkner writes this correspondence into the story of Sutpen, who appears in Mississippi like humankind without apparent origin or purpose. His earliest childhood memories are of a prelapsarian primitive paradise, where "the land belonged to anybody and everybody" and where the country was not "all divided and fixed and neat with a people living on it all divided and fixed and neat because of what color their skins happened to be and what they happened to own" (179). His family's move down from the mountains to Tidewater Virginia parallels the rise of human civilization, "doggeries and taverns now become hamlets, hamlets now become villages, villages now towns and the country flattened out now with good roads and fields and niggers working in the fields while white men sat fine horses and watched them" (182). Most significant, Sutpen's birth of consciousness occurs with the confrontation of his ape-like self in the shape of the master's well-dressed "monkey nigger" butler. That encounter made him look at himself for the first time, and look at his family, and look into his past to see his kind "as cattle, creatures heavy and without grace, brutely evacuated into a world without hope or purpose for them, who would in turn spawn with brutish and vicious prolixity, populate, double treble and compound" (190). A better, more concise literary rendering of the collective reaction to Darwin's *Origin* I cannot presently cite.

The comparison of Sutpen's two halves with Stevenson's "Dr. Jekyll and Mr. Hyde" (1886) brings me to Gillian Beer's discussions of Darwin's legacy in British Victorian fiction. Yes, Faulkner read and reread Conrad and Dickens, and critics have repeatedly noted their influence. But his connection to Victorian writers runs deep and has a major source in evolutionary theory. The ontogeny-phylogeny correspondence, in which man's animal side lurks just under the surface, appears in such works as *Jekyll and Hyde* and H.G. Wells's *The Island of Doctor Moreau* (1896). "In Wells's novel," writes Beer, "the forced hybrids between animals and humans, bred in pain, living in slavery, raise the question of what's human not only by their appearance but by the light they cast on the 'inhumanity' of the experimenters" (1992, 14). Similarly in Faulkner's novel, we are disturbed as much by Sutpen's ruthless treatment of others resulting from his supremely civilized ratiocination as by the deliberate and critical presentation of historical racism by calling African Americans "beastly." The emergence of detective fiction too coincided with post-Darwin literary trends, according to Beer: "Words like *traces* and *decipherment*"—words with a significant role in *Absalom, Absalom!*—"become central to geology, evolutionary theory, and fictional narrative at this time" (1992, 9; 1989, 16).

Beer also refers to Jekyll and Hyde as "an extraordinarily masculinist story: women exist here only on its peripheries, as gloomy housekeeper, lost kin. Self-birthing becomes nightmare" (1992, 38) in a passage that again could summarize Thomas Sutpen's tale. This masculinist aspect derives, Beer argues, from "the threat of matrilineal descent" projected onto evolutionary theory, in which "a female creature would suggest the possibility of reproduction, and raise the suspicion that the missing link might already have been assimilated into an inheritance" (27, 23). In terms of *Absalom, Absalom!*, the lack of mothers in the text observed by Joseph Boone and others also speaks to this threat, especially as it "is complemented by the quite visible, disruptive rise of black sons" (Boone 1997, 1079). Unlike the father, the mother cannot escape acknowledging her connection to a mixed-race child; whereas he can deny it, her body signifies the interracial sexual act. In evolutionary terms, the ability to copulate and reproduce signals that the two parents belong to the same species. The woman's body thus signifies both black and white as human, and threatens the very categorization that permits slavery and sustains the South's way of life. Faulkner communicates this threat through the sexually charged and evolutionary language describing Charles Bon's fractionally black morganatic wife, who exhibits a "female principle which existed, queenly and complete, in the hot equatorial groin of the world long before that white one . . . came down from trees and lost its hair and bleached out" (1990a, 92). And it is Clytie, both *woman* and *black* (of a race "older and purer" than Rosa's [110]), who burns down Sutpen's house.[4]

I should note here that Freud's revolutionary ideas were made possible because, Beer contends, he used "an evolutionist" approach by focusing psychological issues upon origins "always antecedent to language and consciousness"(1989, 164–65; see also 1992, 39). For Beer, the ontogeny-phylogeny correspondence was fundamental to Freud's new psychology. The idea of the beast within, tamed by the civilized part of the mind but nonetheless there, repressed and waiting to return, contributed directly to Freud's division of the psyche into its conscious and unconscious aspects; his concepts of the id and superego echo popularized Darwinism. In other words, Darwin made Freud (and Jung) possible, and consequently made possible Freudian and post-Freudian (Lacanian) interpretations of literature—in our case, the abundance of such interpretations of *Absalom, Absalom!*

II

I hope that I have shown that though the novel's language and surface themes of devolution, determinism, and social shaping of the individual come however directly or indirectly from naturalism, Faulkner's concerns, structure, and influences go beyond naturalism's rather simple portrait of humanity. We can, for example, constructively compare his characters' obsession with history and its continued presence with J. A. Symonds' 1890 description of Darwin and Spencer as turning our conception of the universe into an ongoing process: "No other system has so vigorously enforced the truth that it is impossible to isolate phenomena from their antecedents and their consequents" (Pizer 1976, 80). This impossibility of causal isolation Faulkner weaves into the novel's very structure. Unlike a detective story's or orphan story's linear search for the key clue and origin, two genres Beer sees as direct descendents of Darwin's impact, Faulkner rejects the simple "missing link" plot in recognition of a vastly more complex and naturally accurate structure. Darwin knew that no one race "can be traced back to any one pair of progenitors" (1989, 653). In the evolutionary process species do not simply branch: they branch, branch again, rejoin, refuse, branch again, the different lines always merging with other lines as new lines shoot out, the whole resembling an impenetrable mess. Most often the novel's criticism focuses on the four primary narrators as the story's progenitors: Rosa, Mr. Compson, Quentin, and Shreve. But the story also bears vestiges of Sutpen's own voice, General Compson's voice, and subtle vestiges of more attenuated sources that are present nonetheless: the townspeople whose gossip Rosa invokes, biblical language, and imagistic strains grafted from two Victorians, Oscar Wilde and Aubrey Beardsley (Faulkner 1990a, 157). The contributors to Sutpen's evolving story are as countless and indistinguishable as any living creature's particular ancestors. Indeed untangling Faulkner's narrative, figuring out who is speaking or thinking, and getting the facts straight is so difficult that as late as 1994 Nancy Batty very capably argued that Bon killed Henry, not vice-versa, and that in 1909 Rosa and Quentin discovered Bon, not Henry, hiding in the decaying Sutpen mansion. "There is no point of origin" to Faulkner's characters' history, writes Richard King, "only infinite regress" (1995, 32).

Boone has characterized Sutpen's evolving story as refusing to die, and Philip Cohen has referred to the "obsessive, violent, and ruthless self-assertion" of the story's several versions (1997, 152). No doubt the story evolves, changes, and adapts according to the scene and interests of each telling. Alexandre Leupin correctly observes that "the genealogical obsession of the characters of the novel is a generalized metaphor for the transmission of the story itself" (1990, 227). The selfish gene here involves not Sutpen's DNA but Sutpen's story; human inheritance is as much psychic as it is genetic. Evolution's primary two shaping forces, direct inheritance and environment, Faulkner builds into the evolving story's fate—twice he refers to his characters' "heredity and training" (1990a, 272, 274). Quentin knows Sutpen's story from direct transmission and from having "breathed the same air in which the church bells had rung on that Sunday morning in 1833 [when Sutpen arrived in town] (and, on Sundays, heard even one of the three original bells in the same steeple where descendants of the same pigeons strutted and crooned)" (23). Even at Harvard the conversation occurs with "no listener" and "no talker either" (228), again stressing the indirect shaping over direct transmission. Faulkner makes the analogy between adapting story and adapting organisms explicit in that Harvard room: "what faces and what names they called themselves [Quentin and Shreve and Henry and Bon] and were called by [did not matter] so long as the blood coursed" (237). The marriage trope and sexually suggestive language around Quentin and Shreve may serve less to suggest suppressed homoeroticism than to underscore the genealogy-story metaphor. Quentin and Shreve cannot physically continue a family line, but oh how they can further a story.

The novel's evolving narrative line suggests, then, a direct correspondence with the evolutionary process whereby genetic transmission through sexual intercourse becomes narrative transmission through conversational intercourse. In other words, the correspondence entirely depends upon Faulkner's structure as a dramatization of a series of oral storytelling episodes. Like the unnamed slaves upon whose backs Sutpen's dynasty rests, the unnamed narrative tradition upon whose structure the novel rests is that of African American oral storytelling—the historically less literate, more primarily oral segment of Southern culture. And even if both blacks and whites told (and tell) stories, the former, not the latter, has more deliberately built their storytelling tradition into their printed literature. What is *Absalom, Absalom!* if not what Henry Louis Gates, Jr. has dubbed a "speakerly text," a text that speaks?

Toni Morrison understands "the aesthetic tradition of Afro-American culture," which she aspires to translate into print, as characterized by

> antiphony, the group nature of art, its functionality, its improvisational nature, its relationship to audience performance, the critical voice which upholds tradition and communal values and which also provides occasion for an individual to transcend and/or defy group restrictions. . . . If my work is to be functional to the group (to the village, as it were) then it must bear witness and identify that which is useful from the past and that which ought to be discarded; it must make it possible to prepare for the present and live it out. (1996, 213)

Morrison's storytelling function focuses not on preserving the past perfectly (an impossible task anyway), but on using it to assist living in the present. She allows us to understand better how Shreve's playful variations on the past are infinitely more healthy than Quentin's absorption in it.

Walter Ong provides a more formally critical statement of this function: "Oral societies live very much in a present which keeps itself in equilibrium or homeostasis by sloughing off memories which no longer have present relevance" (1998, 46). In *Orality and Literacy*, Ong details the characteristics of the narratives of primary oral cultures. The problems of speaking from memory and of audience reception require linguistic, syntactic, thematic, and plot element repetition. Vivid scenes of violence, legendary and colorful Sutpen-like "heavy" characters, and formulaic groupings like the many threesomes in *Absalom, Absalom!* serve as mnemonic aids. Such devices, however, do not mean that oral stories stay static. Though the essential elements remain, an oral tale has as many versions as the number of times it is told, and adaptive improvements tend to be perpetuated. Ong's description of "the epic poet's disregard for temporal sequence" in ancient Greek oral narrative reads like a recipe for *Absalom, Absalom!*:

> The poet will report a situation and only much later explain, often in detail, how it came to be. . . . What made a good epic poet was . . . first, tacit acceptance of the fact that episodic structure was the only way and the totally natural way of imagining and handling lengthy narrative, and, second, possession of supreme skill in managing flashbacks and other episodic techniques. (142, 144).

Even the novel's biblical language, as text-based as it appears, comes from the Bible's "orally constituted sensibility and tradition" (99) and reflects the Southern black community's piety in stark contrast with the empty white Christianity of the Sutpens and Compsons. Finally, and as if analyzing Quentin's problem, Ong writes that "there is no way to refute the world of primary orality. All you can do is walk away from it into literacy" (53), yet Quentin can't even do that, can't stop hearing the manifold voices and listening to them merge with his own until he no longer knows where they end and he begins. He needs the model of functional memory described by Morrison and Ong to enable him to live in the present.

III

Reading *Absalom, Absalom!* in a Darwinian and naturalist context leads me to three conclusions the novel ultimately proposes: about Quentin's suicide, about memory, and about miscegenation. First, that Quentin's death informs this novel, that he does not survive much beyond the novel's present, argues for (in Darwinian terms) a fundamental *unfitness* of character. Devolution in *Absalom, Absalom!* does not confine itself to Sutpen's family, which may hold Quentin's fascination as a reflection of his own family's ongoing demise. Mr. Compson, speaking of the "bug-fouled" electric light by which he and Quentin sit on the porch in the dark, speculates that "man had to invent [it] to his need since, relieved of the onus of sweating to live, he is apparently reverting (or evolving) back into a nocturnal animal," and in that very light Mr. Compson's hand appears "almost as dark as a negro's against his linen leg" (Faulkner 1990a, 71). So too Quentin and Shreve sit talking at Harvard, "in the cold room . . . dedicated to that best of ratiocination which after all was a good deal like Sutpen's morality" (225). Shreve with shirt off, battling this nearly interchangeable other for narrative control, reminds this reader of shirtless Sutpen wrestling in the barn with a nearly interchangeable slave. Shreve of course survives this storytelling in a way Quentin does not. Perhaps Quentin Compson is the character most caught in naturalism's trap, with the novels that feature him dramatizing its primary forces, heredity in *The Sound and the Fury*, and his Southern environment in *Absalom, Absalom!* (both novels end with the bellowing of devolved idiots, Benjy Compson and Jim Bond)[5]—though this dichotomy misrepresents the dominance of environment over Quentin in both novels. Quentin fails to thrive because he cannot adapt to his Southern heritage.

Like Sutpen, Quentin tries to wrestle the South, to control it and make it his own. Whereas Sutpen, however, frames the South in his own survivalist mentality and uses southern culture in a very limited and directed manner, trying to adapt it entirely to himself, Quentin tries to control the South by impossibly synthesizing it *in its entirety* into his limited mental frame. He wants all the voices to resolve, almost formulaically; he cannot handle the inherent contradictions, competitions, and infinite complexities. Neither Sutpen nor Quentin successfully balances personal needs with the environment's demands, and thus they fail to thrive, fail to survive. Their failures—and Rosa's too—Faulkner also links to failures in their concepts of memory.

Memory's connection to naturalism, or rather to the ideology behind naturalism, involves what Beer calls "the moral duty to recall and connect" (1989, 15). Post-Darwin Victorians, now suddenly aware "of the vigorous life that long preceded human memory . . . were made preternaturally sensitive to the processes of forgetting and to the extent of what has been forgotten" (17). Thus the rise and popularity of detective fiction are in its quest to remember what happened and thus Freud's moral imperative is for all of us "to restore memories and by such recuperation to make them part of accountable consciousness" (15). Forgetting, one of the most natural and common of human mental activities, had become stigmatized. But in Darwinian terms, human memory, like any operation of any organism, serves not its past but its present and future, through adaptation (see Eakin 1999, 108).

Absalom, Absalom! very much proposes a theory of memory. Sutpen refuses memory by denying his past and ignoring the passage of time in his effort to produce an heir. Having returned from the war, he acts as though "*there might not have been any war at all*" (130), and when he tries for the third time to produce an heir, he slips his last wife Ellen's ring onto her sister Rosa's finger "*as though in the restoration of that ring to a living finger he had turned all time back twenty years and stopped it, frozen it*" (Faulkner 1990a, 133). Beside Sutpen's *amemory* sits Rosa's *monomemory*, the characteristic she shares with her Victorian sister, Dickens's Miss Havisham, which freezes both women in the past and prevents them from any possible psychic development. Quentin's *panmemory* attempts to assimilate all the voices, of the living and the dead, into his own consciousness—a disastrous project contributing to his instability and suicide. He suffers from an overabundant post-Darwin moral duty to recall and preserve. The solution the novel proposes is—to use Morrison's phrase—the *rememory* of oral storytelling. For our psyche to survive its own evolution, we must submit memory to the natural selection process. We cannot force our memories onto our progeny, nor save all our ancestors' memories. The fittest will survive us. As Gloria Naylor's Mama Day recommends, don't fight it; the recollections will end, "at least, in the front part of the mind"; and when they are completely lost, "it won't be the world as we know it no way—and so no need for the memory" (1988, 111).

Finally, this insistence on a kind of memory essentially derived from the Southern African American oral tradition and the embodiment of that tradition in the novel's structure allow Faulkner to comment obliquely on racial miscegenation by means of narrative miscegenation. Darwin's profound influence on literary criticism at the end of the nineteenth century (especially as disseminated by Herbert Spencer) led to genres being treated like biological species such that at least one critic wrote about "a struggle for existence among genres" (Pizer 1976, 82). A century after Darwin and Spencer, in "The Law of Genre," Jacques Derrida in naturalist terms repeatedly expresses this law—that "genres are not to be mixed" (1981, 51). He remarks that genres have been treated in a system akin to "race, familial membership, [and] classificatory genealogy" whereby to mix genres is, by convention, to "risk impurity, anomaly, or monstrosity" (57, 53). It is to commit a kind of miscegenation. Yet his essay finally finds such intermixing inevitable, an inevitability he calls "the law of the law of genre" (55).

The narrative structure of *Absalom, Absalom!* can be viewed as a "cross-breed" of several literary forms. Critics have noted many: the epic, the romance, the Southern Gothic, the detective story, the Dickens orphan tale, etc. If the admixture of so many genres makes a narrative monstrosity of this modern American novel—a generic "perversion," "deformation," or "degenerescence," to use some of Derrida's terms (1981, 53)—it also makes it one of the most sophisticated of its time, its success defying the taboo against mixing genres and by analogy the taboo against mixing races. This essay has focused on two genres: the naturalist novel, which we can conceive as primarily a European-American form, and the oral tale, generally associated in the American South with black culture. The novel's metanaturalism serves to invoke the naturalist novel as a prose genre, to deconstruct the very notion of strict genre categorization by interweaving it with other genres, and to use this generic deconstruction to challenge strict racial categorizations. The embedding of an African American traditional form in the novel's structure dramatically undermines the story's prohibition of white-black sexual relations. In fact, as Faulkner shapes *Absalom, Absalom!*, oral storytelling in the novel finally becomes a redemptive form of memory, adaptation, and survival, so that—as with Dilsey's section in *The Sound and the Fury*,[6] the Compson novel informing this one—Faulkner locates his solution to the debilitating aspects of southern culture by affirming the South's African American heritage.

Notes

Much of the material in this paper appeared in a longer version entitled "Narrative Miscegenation: *Absalom, Absalom!* as Naturalist Novel, Auto/Biography, and African-American Oral Story." *JNT: Journal of Narrative Theory* 3 no. 2 (Summer 2001): 155–79.

1. Lind, Mortimer, and Colatrella, for example, do not treat this novel. Dana Medoro presented an unpublished paper on *Absalom, Absalom!* and evolution-

ary theory at the University of Louisville's 20th Century Literature Conference, 26 February 2000, where I presented a portion of this paper the same day. Medoro contrasted the destructive patriarchal discourse of Sutpen's "might makes right" doctrine with the more integrative matriarchal way of the novel's women characters. For Medoro, Faulkner's references to blood point to menstruation and thus to the matriarchal.

2. According to Ragan, 1987, Faulkner most likely conceived of *Absalom* in 1933, while working on the first volume of the Snopes trilogy *(The Hamlet)* (1), which Mortimer believes is a Darwin-inspired project and even makes "the obvious pairing" of Snopes and Scopes (199). Mortimer casually notes how Faulkner associates Sutpen as he does the Snopses with animal imagery (189) but does not otherwise elaborate on a possible Darwinian paradigm in *Absalom, Absalom!*

3. Like the architect, Charles Bon is made French through his New Orleans home. Charles in fact outdoes Sutpen in his detachment, calculation, and stylish walk such that Henry "apes" Charles every chance he gets. But for that drop of black blood, Charles of all of the children would have best carried Sutpen's genetic heritage. Indeed even Judy has more Sutpen in her than Henry has; were she and Charles to marry, they would surely bear their father's best hope. That they aren't allowed to marry reiterates the point that Faulkner makes with Sutpen: when human rules try to supercede natural laws, doom results.

4. Some purely coincidental notes: Darwin (b.1809) and Sutpen (b.1807?) were born within a year or two of one another, certainly within Sutpen's margin of accuracy. Their mothers died the same year (1817). Sutpen's project Henry was born in 1839, the year Darwin began *Origin*, and Henry met Charles, arguably the year Sutpen's design began to come undone, in 1859—the year Darwin published *Origin*. Finally, a passage from Darwin's autobiography recalls Sutpen's Spartan lifestyle and attitude: "I have said that in one respect my mind has changed during the last twenty or thirty years. Up to the age of thirty, [poetry, plays, art, and music] gave me great pleasure. . . . But now for many years I cannot endure to read a line of poetry. . . . My mind seems to have become a machine for grinding general laws out of large collections of facts, but why this should have caused the atrophy of that part of the brain alone, on which the higher tastes depend, I cannot conceive. . . . The loss of these tastes is a loss of happiness, and may possibly be injurious to the intellect, and more probably to the moral character, by enfeebling the emotional part of our nature" (1958, 53–54).

5. For Mitchell, heredity and the environment render the self as an illusion (1988, 531). Quentin's physical body torments him as evidence of selfhood and denial of that self; his obsession with virginity is an obsession to deny the necessity of sexual maturation and reproduction, to deny his body's fatalistic demands—including death, that loss of self and by implication the possibility of self as only a temporary illusion, which the sexual impulse ultimately signifies.

6. In that novel the redeeming presence is the preacher who visits Dilsey's church, a man with "a wizened black face like a small, aged monkey," who initially speaks "like a white man" and to whom the congregation listens "as they would have to a monkey talking." As he continues to talk his language gradually changes, "became negroid." This man, with the body and face of a monkey, can discourse in the language of both white and black, and in the process ascends to a level of spirituality unknown in either novel (1990b, 293–97).

References

Batty, Nancy E. 1994. "The Riddle of Absalom, Absalom: Looking at the Wrong Blackbird?" *Mississippi Quarterly* 47 no. 3 (Summer): 461–89.

Beer, Gillian. 1989. *Arguing with the Past: Essays in Narrative from Woolf to Sidney*. London: Routledge.

———. 1992. *Forging the Missing Link: Interdisciplinary Stories*. Inaugural Lecture, 18 November 1991. Cambridge, Mass.: Cambridge University Press.

Blotner, Joseph. 1974. *Faulkner: A Biography*. One-volume edition. New York: Random House.

Boone, Joseph A. 1997. "Creation by the Father's Fiat: Paternal Narrative, Sexual Anxiety, and the Deauthorizing Designs of Absalom, Absalom!" In *Feminisms: An Anthology of Literary Theory and Criticism*, edited by Robyn R. Warhol and Diane Price Herndl. New Brunswick, N.J.: Rutgers University Press.

Cohen, Philip. 1997. "Faulkner." *American Literary Scholarship: An Annual 1995*, edited by Gary Scharnhorst. Durham, N.C.: Duke University Press.

Colatrella, Carol. 1990. *Evolution, Sacrifice, and Narrative: Balzac, Zola, and Faulkner*. New York and London: Garland Publishing.

Cowley, Malcolm. 1956. "Naturalism in American Literature." In *Evolutionary Thought in America*, edited by Stow Persons. New York: George Braziller.

Darwin, Charles S. [1929] 1958. *The Autobiography of Charles Darwin and Selected Letters*. Reprint, edited by Francis Darwin. New York: Dover.

———. [1871] 1989. *The Descent of Man, and Selection in Relation to Sex*. Reprint, *The Works of Charles Darwin*. Vol. 2, edited by Paul H. Barrett and R. B. Freeman. Advisor, Peter Gautrey. New York: New York University Press.

Derrida, Jacques. 1981. "The Law of Genre." In *On Narrative*, edited by W. J. T. Mitchell. Chicago: University of Chicago Press.

Donaldson, Susan V. 1998. *Competing Voices: The American Novel 1865–1914*. Twayne's Critical History of the Novel. New York: Twayne.

Eakin, Paul John. 1992. *Touching the World: Reference in Autobiography*. Princeton, N.J.: Princeton University Press.

———. 1999. *How Our Lives Become Stories: Making Selves*. Ithaca, N.Y.: Cornell University Press.

Faulkner, William. [1936] 1990a. *Absalom, Absalom!* Reprint, New York: Vintage International.

———. [1929] 1990b. *The Sound and the Fury*. Reprint, New York: Vintage International.

Hobson, Fred C., Jr. 1974. *Serpent in Eden: H. L. Mencken and the South*. Chapel Hill: University of North Carolina Press.

Irwin, John T. 1996. *Doubling and Incest/Repetition and Revenge: A Speculative Reading of Faulkner*. Expanded edition. Baltimore, Md.: Johns Hopkins University Press.

King, Richard H. 1995. "Faulkner, Ideology, and Narrative." In *Faulkner and Ideology: Faulkner and Yoknapatawpha 1992*, edited by Donald M. Kartiganer and Ann J. Abadie. Jackson: University Press of Mississippi.

Leupin, Alexandre. 1990. "*Absalom, Absalom!*: The Outrage of Writing." In *Southern Literature and Literary Theory*, edited by Jefferson Humphries. Athens: University of Georgia Press.

Lind, Ilse Dusoir. 1980. "Faulkner and Nature." In *Faulkner Studies I*, edited by Barnett Guttenberg. Miami, Fla.: University of Miami Press.

Morrison, Toni. 1996. "Memory, Creation, and Writing." In *The Anatomy of Memory: An Anthology*, edited by James McConkey. New York: Oxford University Press.

Mortimer, Gail. 1986. "Evolutionary Theory in Faulkner's Snopes Trilogy." *Rocky Mountain Review of Language and Literature* 40 no. 4: 187–202.

Mitchell, Lee Clark. 1988. "Naturalism and Languages of Determinism." *Columbia Literary History of the United States*. Emory Elliot, general editor. New York: Columbia University Press.

Naylor, Gloria. 1988. *Mama Day*. New York: Vintage.

Ong, Walter J. 1998. *Orality and Literacy: The Technologizing of the Word*. London: Routledge.

Pizer, Donald. 1976. *Realism and Naturalism in Nineteenth-Century American Literature*. Preface by Harry T. Moore. New York: Russell and Russell.

Ragan, David Paul. 1987. *William Faulkner's* Absalom, Absalom!: *A Critical Study*. Ann Arbor, Mich.: UMI Research Press.

Russett, Cynthia Edge. 1976. *Darwin in America: The Intellectual Response 1865–1912*. San Francisco, Calif.: W. H. Freeman.

II
Eugenics in American Literature

Eugenics and the Fiction of Pauline Hopkins

John Nickel

"[O]UR PEOPLE ARE IMPROVING IN THEIR DRESS, IN THEIR LOOKS and in their manners," states an African American character in Pauline Hopkins's first of four novels, *Contending Forces*, published in 1900 (1988a, 110). Faced with the fierce post-Reconstruction attacks on African American civil rights and humanity, many African American writers sought to demonstrate progress of the black race. What is striking about this particular statement is its assertion that African Americans' physical features ("their looks") are improving, implying that the race's progress is not only cultural but also biological. While Hopkins agreed with the majority of contemporary black writers that educational and moral progress was important to racial uplift, she also prescribed another remedy. Influenced by the eugenic belief in the "improvement of the human race through better breeding," to quote the leading United States eugenist Charles Davenport in 1911, Hopkins advocated that African Americans' genetic improvement was necessary for racial advancement, and dependent on their marital choices (1913, 1). Calling for the commingling of white and black racial lines, Hopkins asserted that it would produce a genetically superior race and eventually lead to the amelioration of African Americans' political and social conditions.

Hopkins's promotion of eugenics for racial uplift, however, was problematic. Given the racial, gender, and class prejudices of contemporary eugenics, her assimilationist agenda had the unavoidable effect of reinforcing its demeaning logic. The fact that a radical writer such as Hopkins appropriated eugenic tenets as one of the means for racial improvement points to the desperate situation that African Americans faced in the late-nineteenth and early-twentieth centuries. While segregation and disenfranchisement substantially blocked public avenues for social change, Hopkins, in order to construct a program for racial progress, turned to intimate areas of life—marriage and reproduction—that could be selfregulated by African Americans. Her fiction manifests the power of coeval scientific discourses to set the framework and terms for many debates over racial and social equality. Surveying Hopkins's fiction while focusing on her eugenic agenda will enable us to study the employment of science to resist racism, as well as explore the complex and intersecting constructions of race, gender, and class at the turn of the twentieth century.

Hopkins's eugenic vision must be considered in light of her larger social and political response to contemporary race problems. Without question, Hopkins was a literary pioneer who viewed her fiction as a way to combat stereotypes of black Americans. "In giving this little romance expression in print," she wrote in the preface to *Contending Forces*, "I am not actuated by a desire for notoriety or for profit, but to do all I can to raise the stigma of degradation from my race" (1988a, 13). Calling attention to racist practices such as segregation and lynching, Hopkins's work as a fiction writer, journalist, and editor of *The Colored American Magazine* encouraged political protest among her readership. The radical nature of Hopkins's political position is evident from her leaving *The Colored American Magazine* after its change of ownership in 1904, when the accomodationist Booker T. Washington became the periodical's primary financial supporter. Writing in *The Crisis,* W. E. B. Du Bois later recalled Hopkins's departure: "It was suggested to the editor . . . that her attitude was not conciliatory enough" (Johnson and Johnson 1979, 9).

During a period that historian Rayford Logan describes as the "nadir" of African American history because of entrenched and violent white racism, Hopkins felt that racial amalgamation would ultimately resolve racial conflicts (Logan 1965, 88). By eliminating racial differences, biological race assimilation would overcome deep-seated prejudices between the two races and put an end to racist practices such as segregation. Expressing this view, one character states in Hopkins's second novel, *Hagar's Daughter*: "This idea of race separation is carried to an extreme point and will, in time, kill itself. Amalgamation has taken place; it will continue, and no finite power can stop it" (Hopkins 1988b, 270). In opposition to contemporary arguments against racial mixing by white and black nationalists, Hopkins suggests that the blood of Anglo and African Americans is already hopelessly intermixed and that racial homogeneity therefore is unrealistic.

Hopkins's melodramatic and complex plots dramatize the extent to which racial lines have been commingled.[1] Similar to the main characters in Frances Harper's *Minnie's Sacrifice* (1869), the protagonists in Hopkins's works often are not fully aware of their true genealogy at the beginning of the narrative, and, during the course of the story, they find out they have black heritage and have been passing unwittingly as white.

Hagar's Daughter, which was published serially in *The Colored American Magazine* in 1901 and 1902, follows this plot structure: the title character learns of her African descent after she has married a white, southern aristocrat, Ellis, and given birth to a daughter. Employing several sensational plot devices, Hopkins has Hagar separate from Ellis, pass for white for twenty years, and then reunite with him and quietly remarry. Hagar's passing, both unknowingly and intentionally, plays upon white fears that persons of mixed racial descent have secretly entered the white race and reinforces the inability of society to keep the racial lines distinct. Discussing persons of "mixed-blood," one character in the novel states: "black blood is everywhere—in society and out, and in our own [white] families even. . . . We try to stem the tide but I believe it is a hopeless task" (1988b, 160). In the context of the hysteria over racial mixing at the turn of the century, Hopkins offers an effective refutation of the Jim Crow ideology of racial purity.

Hopkins shared Charles Chesnutt's prophecy of the complete union of the black and white racial lines and believed that the new mixed race would be eugenically superior to both the black and white races. Certain traits of African American blood, Hopkins believed, would benefit the Anglo American race. One such trait was virtue, which an African American character in *Contending Forces* states is "an essential attribute peculiar to us—a racial characteristic which is slumbering but not lost" (1988a, 149). Her fiction advocates not the obliteration of black racial traits, but rather a combination of black and white traits. Given the dominant belief that black Americans' heredity was generally inferior to white Americans' heredity and that the inferior traits were largely responsible for the subordinate status of the former, Hopkins suggests that the black race needs white racial traits in order to improve the race's position within society.

Hopkins articulates her position in a narratorial intrusion in *Contending Forces*, in which, similar to contemporary writings on racial mixing, human breeding is compared to plant breeding:

> Why should we wonder or question, then, when we see the steady advance of a race overriding the barriers set by prejudice and injustice? Man has said that from lack of means and social caste the Negro shall remain in a position of serfdom all his days, but the mighty working of cause and effect, the mighty unexpected results of the law of evolution, seem to point to a different solution of the Negro question than any worked out by the most fertile brain of the highly cultured Caucasian. Then again, we do not allow for the infusion of white blood, which became pretty generally distributed in the inferior black race during the existence of slavery. Some of this blood, too, was the best in the country. Combinations of plants, or trees, or of any productive living thing, sometimes generate rare specimens of the plant or tree; why not, then, of the genus homo? Surely the Negro race must be productive of some valuable specimens, if only from the infusion which amalgamation with a superior race must eventually bring. (1988a, 87)

Richard Yarborough, interpreting this passage in his introduction to *Contending Forces*, suggests that the phrase, "the most fertile brain of the highly cultured Caucasian," can be read as sarcastic, given the negative portrayal of white Americans in the novel. As Yarborough notes, however, the subsequent references to "white blood" and "superior race" cannot be easily dismissed (1988a, xxvi).

In novels such as *Contending Forces*, Hopkins addresses her eugenic program to the largely bourgeois, mixed-race, female readership of the black domestic romance genre. Functioning as eugenic conduct books, Hopkins's courtship stories instruct her audience on appropriate selections for marriage. A significant plot conflict in *Contending Forces*, for example, is who the spirited Dora Smith should marry. A mixed-race woman of the middle-class, Dora has inherited eugenically-fit traits from her mother, Mrs. Smith, whose "superior intelligence" and good manners in turn, we learn, came from her white aristocratic ancestors (1988a, 373). Dora is engaged to marry the villain John Langley, who inherited his low "moral nature" and "revengeful trait of character" from the lower-class paternal side of his family—" 'cracker' blood of the lowest type" (21). But the better marital choice would be Arthur Lewis; ironically, given Hopkins's reason for leaving *The Colored American Magazine*, Lewis is modeled after Booker T. Washington and is president of a respected southern technical college. When Dora discovers that Langley has attempted to blackmail her friend Sappho Clark for sexual favors, Dora terminates the engagement immediately and decides to marry Lewis, thus preventing the traits of Langley's inferior "cracker blood" from being passed on to her children. Fortunately for the genetic future of the race, Langley, without ever marrying, dies while searching for gold in the Klondike.

Just as Hopkins used a literary medium (domestic romance novels) that addressed women readership, so did other eugenic reformers, who often published their writings in women's journals. Commenting on the eugenic assumption that women were the proper audience for programs of racial improvement, scientific historian Daniel Kevles writes: "Eugenics, concerned ipso facto with the health and quality of offspring, focused on issues that, by virtue of biology and prevailing middle-class standards, were naturally women's own" (Kevles 1985, 64). For example, in a 1913 article in *Cosmopolitan Magazine*, entitled "Do You Choose Your Children?," Stoddard Goodhue told his women readers that the wrong marital choice could lead to children who are "feeble minded or epileptic or sexually perverted or destined to become insane" (150). A friend of Charles Davenport and a eugenics supporter, President Theodore Roosevelt also entered the fray, as part of his personal campaign to safeguard the nation's genetic makeup and thereby keep the nation strong. Roosevelt delivered a speech before the National Congress of Mothers in 1905, which was printed in *The Ladies Home Journal* under the title, "The American Woman as a Mother." A firm believer in "race suicide," that Anglo Americans were losing ground to the growing populations of African Americans and immigrants, Roosevelt called on white women to have large families. The "average woman," proclaims Roosevelt, should be a "good wife, a good mother, able and willing to perform the first and greatest duty of womanhood, able to bear, and to bring up as they should be brought up, healthy children, sound in body, mind and character, and numerous enough so that the race shall increase and not decrease" (1905, 3–4). The proper selection of a mate, motherhood, devotion to domestic life, and maintenance of family ideals—these were all the obligations of women, according to Roosevelt and other eugenic writers, to ensure the betterment of the (white) race.

The duty of women to their race is the topic of a women's sewing club meeting in *Contending Forces*. The participants have gathered to hear a talk on the subject, "The place which the virtuous woman occupies in the upbuilding of the race," given by Mrs. Willis, a public spokesperson on "the evolution of true womanhood in the work of the 'Woman Question' as embodied in marriage and suffrage" (Hopkins 1988a, 148, 146). Referring to the lineage of Mrs. Willis, Hopkins emphasizes that "a strain of white blood had filtered through the African stream" (145). Described as a "brilliant widow," Mrs. Willis addresses her racial uplift talk to an audience, resembling that of Hopkins's own readership, of mixed-race bourgeois women: Sappho Clark, Dora Smith, and Mrs. Smith, among others (145). Concerned with refuting the stereotype of black women as lecherous and uncontrollably passionate, Mrs. Willis tells her audience that they and their children must counter this stereotype through proper behavior. Practicing Christian self-control, in this case not giving into sexual "temptation," is essential (149).

By using sexuality appropriately and avoiding marriages with debauched men such as John Langley, women of mixed African descent, in Hopkins's thinking, would eventually purify the racial bloodlines of immoral traits. To this effect, Mrs. Willis urges her audience to "hasten the transformation of the body by the nobility of the soul" and "cultivate, while we go about our daily tasks, no matter how inferior they may seem to us, beauty of soul and mind, which being transmitted to our children by the law of heredity, shall improve the race by eliminating *immorality* from our midst and raising *morality* and virtue to their true place" (153). Echoing other passages in Hopkins's fiction, Mrs. Willis states that "the black race on this continent has developed into a race of mulattoes" and that God has prepared a "niche" for the race that corresponds to the race's "innate fitness" (151, 152). In situating these mixed-race eugenic views in the context of the lively discussion of the sewing circle, Hopkins gives her readers a sense of collective mission, of women working together for race betterment and social change.

As in earlier sentimental women's fiction such as *Uncle Tom's Cabin*, the domestic sphere in Hopkins's works is transformed into an important site of social and political change. According to Hopkins, racial progress will occur in the domestic realm—specifically and even more privately, in the institution of marriage and reproduction. Black women, then, are not peripheral but central to racial uplift: they are the guardians of the race's genes, the controllers of the characteristics to be passed on to future generations. Though Hopkins, like mainstream eugenists, offers women a new avenue for social activism, the role for women is circumscribed within patriarchal ideologies of femininity and female sexuality. Following the Victorian doctrines of "true womanhood," women should practice self-control and chastity and emulate the ideal characteristics and behavior of Hopkins's woman protagonists: "refined, cultured, possessed of all the Christian virtues" (1988b, 62). Women's primary role in Hopkins's uplift program is, in effect, to perform appropriately a biological function, to limit sexuality to reproduction in marriage and transmit favorable heredity to posterity. In keeping with contemporary eugenic literature, the female body becomes an instrument to be controlled for social purposes, a vehicle to be used first and foremost for the service of the race.

As models for Hopkins's intended readers, her mixed-race protagonists carefully negotiate racial and class differences when considering a mate for marriage. The heroines wed white and mixed-race persons but not individuals of unmixed African descent because of their presumed inferior heredity. Unsuitable marital choices also include individuals of the lower class. In her effort to construct a positive future for individuals of biracial heritage, Hopkins relies on widely-held beliefs in class differences. A member of the rising black middle-class, Hopkins posits that black and white racial amalgamation occurred historically with middle- and upper-class whites. This fundamental change in the genetic makeup of the African American race was brought about by slavery, as Hopkins discusses in the short story, "A Dash for Liberty," published in 1901 (1991a). The narrator of this story states that a white grandfather of a beautiful "octoroon" served in the Revolutionary War and both Houses of Congress. "That was nothing, however," the narrator avers, "at a time when the blood of the proudest F. F. V.'s [First Families of Virginia] was freely mingled with that of the African slaves on their plantations. Who wonders that Virginia has produced great men of color from the exbondsmen, or, that illustrious black men proudly point to Virginia as a birthplace?" (94). Unlike writers in the plantation tradition, such as Thomas Nelson Page, who depicted nostalgic, picturesque scenes of slaveholding estates as places where whites and blacks respected their unequal social positions, Hopkins saw only one positive effect of slavery: interracial unions among African Americans and elite whites. The narrator continues, "Posterity rises to the plane that their ancestors bequeath, and the most refined, the wealthiest and the most intellectual whites of that proud State have not hesitated to amalgamate with the Negro" (94).

An extension of Hopkins's expressed view is that the fin de siècle class structure within the African American community reflects the patterns of racial mixing during slavery; in other words, the "illustrious" and "great men of color" owe their class position in part to the "blood" of their white aristocratic ancestors. The implication of this reasoning is that racial amalgamation in the counter-Reconstruction era should occur among the upper classes, black and white, in order to preserve the best blood.

This view is expressed in another short story by Hopkins, "Talma Gordon," published, like "A Dash for Liberty," in *The Colored American Magazine*. In "Talma Gordon," which appeared in 1900 (1991b), the white male members of an elite Boston club are discussing United States imperialist policy when they broach the subject of racial intermixing. To the astonishment of the other members of the club, Dr. Thornton, the story's main character, claims that marriages with non-Caucasians are not only natural but are beneficial when "they possess decent moral and physical perfection, for then we develop a superior being in the progeny born of the intermarriage" (1991b, 51). However, interracial unions among lower-class individuals will not produce this improved human type, suggests Dr. Thornton, who describes miscegenation among all classes as "appalling" (51). The story concludes with one of Hopkins's characteristically melodramatic moments, Dr. Thornton's revelation that he has married the title character, Talma Gordon, a beautiful, culturally refined woman of mixed racial heritage. With the Anglo setting of "Talma Gordon," Hopkins appears to be addressing with this story the minority white readership of *The Colored American Magazine*.

The belief that those best fit for mating were from the upper classes was in keeping with the class bias of contemporary eugenics, whose supporters in the United States and England were mostly educated, middle- to upper-class individuals. A British eugenist later recalled with regret how, in the years before 1914, "social class was sometimes put forward as a criterion of eugenic value; and terms sometimes were used such as 'lower classes,' 'riff raff,' 'dregs,' which seemed to imply a contempt for certain sections of the poor" (Bannister 1979, 177). As in "A Dash for Liberty," the elitism of mixed-race eugenic supporters led them to minimize that racial amalgamation often occurred as a result of the systematic and brutal rapes of slave women by lower-class slave masters.[2] The treatment by Hopkins of racial intermixing during slavery resembled that of another writer at *The Colored American Magazine*, Daniel Murray. In an article entitled, "Race Integrity—How to Preserve It in the South" (1906), Murray attacks racist beliefs that the mulatto is inferior and sterile, criticizing the "old theories" on which these beliefs are based. On the contrary, children of "mixed-blood" are superior to black or white children, Murray maintains, because the first group was likely procreated by black mothers and southern cavaliers, not "some low bred master," and by people who loved each other intensely, defying antimiscegenation laws (Murray 1906, 375). Praising that the United States genetic makeup is gradually becoming biracial, Murray concludes that miscegenation "will in time solve the race question by eliminating the Negro, the new race element filling the void" (377).

Hopkins's class and racial preferences intersect in her treatment of marriage in *Winona*, a novel set in antebellum America and published in serial form in *The Colored American Magazine* in 1902. The romantic tension of the plot centers around three characters: Judah, a character of unmixed African descent; Winona, a woman of mixed racial heritage; and Warren, a white English aristocrat. Winona and the orphaned Judah have been raised in a remote setting by Winona's white aristocratic father, whose murder leads to their enslavement. After they escape the slave plantation with Warren's aid, Judah considers expressing his love to Winona, but he is intimidated by her superior social status, exemplified in her refinement and grace. "Unconsciously," the narrator states, "he was groping for the solution of the great question of social equality." "But," the narrator continues,

> is there such a thing as social equality? There is such a thing as the affinity of souls, congenial spirits, and good fellowship; but social equality does not exist because it is an artificial barrier which nature is constantly putting at naught by the most incongruous happenings. Who is my social equal? He whose society affords the greatest pleasure, whose tastes are congenial, and who is my brother in the spirit of the scriptural text, be he white or black, bond or free, rich or poor. (1988d, 377)

This narratorial intrusion implies that social distinctions are natural and should be defined by similar "tastes," as opposed to racial or material differences. Overcoming his inhibitions, Judah expresses his love to Winona, who reveals her love instead for Warren (378). Warren, not Judah, is Winona's social equal.

Winona, however, forsakes marriage because American racial prejudices will prevent her from marrying someone from within her father's aristocratic class. Later in the narrative, she declares, "I cannot marry out of the class of my father. . . . It follows, then, that I shall never marry" (406). Winona's situation is resolved when she and Warren, along with Judah, move to England, where, uninhibited by the racism in the United States, Winona and Warren plan to marry. England functions in the denouement as tabula rasa, an absent space where the characters can create a new life without racial prejudice. Like William Dean Howells' *An Imperative Duty* (1893), in which the white bourgeois hero marries a mixed-race bourgeois heroine, *Winona* suggests that marriage between mixed-race individuals and whites is permissible, but only among the upper classes.

Hopkins's advocacy of creating a mixed race through appropriate marriages, as we see in *Winona*, stems from her belief in a divinely ordained progression of the human race. Hopkins elaborates on her theory of human progress in a 1905 pamphlet entitled, *A Primer of Facts Pertaining to the Early Greatness of the African Race and the Possibility of Restoration by Its Descendants—with Epilogue*. God's purpose in human creation, according to this short treatise, is to promote God's glory through the development and improvement of human civilization. Arguing that Western civilization originated in mighty, ancient Ethiopian kingdoms, Hopkins provides a historical record of the ancient Africans' achievements and laments the subjugated status of modern-day persons of African descent in the United States. Individuals of Ethiopian heritage, though, will again rise to positions of power, suggests Hopkins, who invokes Ethiopianism, the prophecy of the future glory of Ethiopians based on the Biblical verse that "Ethiopia shall soon stretch forth her hands to God."

The treatise also describes the origins of the races, citing the account of creation in *Genesis*—all the races came from Adam, the "parent stock," and then separated after the destruction of the Tower of Babel. With a complexion that "must have been clay or yellow," Adam is depicted as a racially composite prototype whose descendants "were of one race and complexion" (Hopkins 1905, 5). The pamphlet ends with a eugenic vision of a new mixed race in the United States, stating that "Anglo-Saxon blood is already hopelessly perverted, with that of other races, and in most cases to its great gain" and that "amalgamation may produce a race that will gradually supercede the present dominant factors in the government of this Republic" (28–29). According to the logic of that conclusion, the growing mixed race in the United States will fulfill God's mission for human progression by reuniting the blood lines that were separated in ancient times. Joining eugenics with Ethiopianism and Biblical creationism, Hopkins suggests that the mixed race in the United States will come to resemble Adam and reestablish the greatness of the early African race.

Many of the themes of *A Primer*—the concern with racial "blood" lines, the cradle of Western civilization in Africa, and the early and approaching glory of individuals of African heritage—appear in Hopkins' last novel, *Of One Blood*, which was published serially in *The Colored American Magazine* from 1902 to 1903. Though written in the adventure tale genre, *Of One Blood* resembles Hopkins's other major works in that its plot conflict resolves in marriage. The male hero of the novel, Reuel Briggs, is a Harvard-educated, mixed-race doctor who intentionally passes as white and who has phenomenal powers: clairvoyance and an ability to revive recently deceased people. Reuel marries a beautiful, near-white woman of mixed racial heritage, Dianthe Lusk—who, the narrator states, "was not in any way the preconceived idea of a Negro"—but shortly thereafter he takes a medical appointment by himself in Africa (Hopkins 1988c, 453). On an expedition there, Reuel discovers an ancient Ethiopian kingdom, Telassar, which he learns was the source of Western civilization and is still inhabited by an ancient race who lived in Babel before its destruction. Though isolated from modern civilization, the inhabitants of Telassar are highly advanced people, "preserved specimens of the highest attainments the world knew in ancient days" (551). When a lotus birthmark on Reuel's breast proves that he is the long-awaited King of Telassar, who will restore the kingdom to greatness, Reuel embraces his black heritage. After his wife Dianthe dies tragically, Reuel decides to live in the ancient kingdom permanently and to marry Telassar's Queen Candace.

Pointing to the description of Queen Candace's beauty—"long, jet-black hair and totally free, covered her shoulders . . . a warm bronze complexion; thick black eyebrows, great black eyes . . . a delicate nose with quivering nostrils" (568–69)—critics have read *Of One Blood* as a fictional representation of "Black is Beautiful."[3] Furthermore, commentators, noting the novel's Ethiopianism, Reuel's acceptance of his black heritage, and his union with Queen Candace, have argued that the work rebukes racial mixing and embraces pure black bloodlines. Jennie Kassanoff maintains that the novel makes a "forceful protest against miscegenation" and "urges the return to the purity of 'Negro blood'" (1996, 168), while Eric Sundquist asserts that Reuel's "white blood and skin" are "[s]heared away in his return to African kinship" and that Reuel escapes the racial mixing that is "destructive of African purity in the United States" (1993, 572). In agreement with these readings, Hazel Carby states that the "establishment of Ethiopia as the source of all civilization . . . allowed Hopkins to eliminate from her fiction any significance attached to Anglo-Saxon heritage on which so many of her stories depend for their denouements" (1987, 158). Such interpretations suggest that, by the time Hopkins wrote her last novel, she had distanced herself from earlier preferences for white racial descent and now celebrated solely black racial descent.

These readings of *Of One Blood* assume that Queen Candace and the other ancient Ethiopians represent individuals of pure or unmixed black racial heritage.[4] The physical descriptions of these characters, however, reveal them to be multiracial. When Reuel first enters the kingdom, he notices that the inhabitants "ranged in complexion from a creamy tint to purest ebony; the long hair which fell on their shoulders, varied in texture from soft, waving curls to the crispness of the most pronounced African type. But the faces into which he gazed were perfect in the cut and outline of every feature; the forms hidden by soft white drapery, Grecian in effect, were athletic and beautifully moulded" (Hopkins 1988c, 545). The "trace of crispness" (560) in the hair of Reuel's Ethiopian teacher, Ai, bears remarkable similarity to the hair of the mixed-race Will Smith in *Contending Forces*—which has "just a tinge of crispness to denote the existence of Negro blood" (Hopkins 1988c, 90). Furthermore, Queen Candace reminds Reuel "strongly" of his mixed-race wife: "the resemblance was so striking that it was painful . . . She was the same height as Dianthe, had the same well-developed shoulders and the same admirable bust" (1988c, 568). Manifestations of Hopkins's belief in "the Ethiopian as the primal race," Queen Candace, Ai, and the other inhabitants of Telassar are portrayed as members of the prototypic mulatto race, the original race from which the modern black and white races derived (521). The novel's final line, a Biblical quotation that appears in many of Hopkins's works, reinforces the point that the races sprang from a common origin: "Of one blood I have made all the races of men" (621).

While Hopkins's fiction never subscribed to "negative eugenics," that is, limiting those considered unfit for reproducing, her works endorsed marriages among the supposedly eugenically fit members of society. *Of One Blood* glorifies not blackness per se but rather mixed racial heritage. Leaders of the ancient world antedating Egyptian civilization, the racially composite people of Ethiopia in the novel are known for the race's luxurious palaces, arts, and occult powers; they are able to know the past, present at faraway places, and future. This "once magnificent race" even possesses a knowledge of science superior to modern times (547). When the modern-day races became distinct after the destruction of Babel, in Hopkins's view, the races declined and lost much of the remarkable abilities of the original people. But, prefiguring the conclusion of *A Primer*, the narrator states that God has created a new mulatto race in the United States: "in His own mysterious way He has united the white race and the black race in this new continent" (607). With his incredible medical skill and mysticism, Reuel resembles his Ethiopian ancestors and exemplifies the powers of the new mixed race, the "modern Ethiopian" (560); his mixed "blood," then, not solely his black "blood," is necessary for him to be King of Telassar.

As a representative of the mixed race, Reuel points to the possibility of future racial unity and of the establishment of a peaceful kingdom like Telassar; yet, his decision to live in Africa permanently because of United States racial strife serves as Hopkins's reminder to her audience that post-Reconstruction society is not yet ready to accept this new race. Similar to the marriage of Dora Smith and Arthur Lewis in *Contending Forces*, *Of One Blood* closes with the union of two mixed-race individuals—an original prototype, Queen Candace, and perhaps the most advanced modern person, Reuel. Thus, Hopkins's final novel is a testimonial to the early greatness of the first multiracial people, to the potential for its restoration by eugenic racial progress in the United States, and to the hope (unfortunately unrealized) for a time when prejudices and injustices against African Americans would be in the distant past.

NOTES

Reprinted by permission of *ATQ, Nineteenth Century American Literature and Culture* 14 no. 1, March 2000, at the University of Rhode Island.

1. On this topic, see Carby's discussion of *Contending Forces* (1987, 140).

2. As Carby reminds us, Hopkins used her fiction (particularly *Contending Forces*) to protest the rape of African American women. In Carby's words, "What Hopkins concentrated on . . . was a representation of the black female body as colonized by white male power and practices" (1987, 143).

3. Claudia Tate suggests that Queen Candace "is the means by which pigment is reintroduced into the royal line, inasmuch as [King] Ergamenes (alias Reuel) lost that trait as a result of miscegenation. Hence, pigment becomes a positive physical attribute measured in terms of feminine beauty more than sixty years prior to the coinage of the slogan 'Black is Beautiful'" (Tate 1985, 65). More ambivalently, Carby states that the "idealization of black beauty in the novel is contradictory and retains classically European pretensions. Hopkins selects black skin, eyes, brows, and "crisp" black hair for praise, but profiles and bone structure remain Athenian. . . . Hopkins sought to externally authenticate her fictional representation of black as beautiful" (1988, xlvi).

4. As the readings indicate, critics have concurred that Hopkins's last novel, *Of One Blood,* promotes the purity of black racial lines. Other interpretations presume that the ancient Ethiopians depicted in the novel are of pure black descent or that Hopkins advocates a return to unmixed black lineage: for example, Ammons 1991, 81–85; Bruce 1989, 153–55; and Gillman 1992, 231–36, and 1996, 57–82.

References

Ammons, Elizabeth. 1991. *Conflicting Stories: American Women Writers at the Turn into the Twentieth Century.* New York: Oxford University Press.

Bannister, Robert C. 1979. *Social Darwinism: Science and Myth in Anglo-American Social Thought.* Philadelphia, Pa.: Temple University Press.

Bruce, Dickson D., Jr. 1989. *Black American Writing from the Nadir: The Evolution of a Literary Tradition, 1877–1915.* Baton Rouge: Louisiana State University Press.

Carby, Hazel V. 1987. *Reconstructing Womanhood: The Emergence of the Afro-American Woman Novelist.* New York: Oxford University Press.

———. 1988. Introduction. In *The Magazine Novels of Pauline Hopkins.* New York: Oxford University Press.

Davenport, Charles. 1913. *Heredity in Relation to Eugenics.* New York: Henry Holt.

Gillman, Susan. 1992. "The Mulatto, Tragic or Triumphant? The Nineteenth-Century American Race Melodrama." In *The Culture of Sentiment: Race, Gender, and Sentimentality in Nineteenth-Century America,* edited by Shirley Samuels. New York: Oxford University Press.

———. 1996. "Pauline Hopkins and the Occult: African-American Revisions of Nineteenth-Century Sciences." *American Literary History* 8: 57–82.

Goodhue, Stoddard. 1913. "Do You Choose Your Children?" *Cosmopolitan Magazine* 55 (July): 148–57.

Hopkins, Pauline E. 1905. *A Primer of Facts Pertaining to the Early Greatness of the African Race and the Possibility of Restoration by Its Descendants—with Epilogue.* Cambridge, Mass.: P. E. Hopkins and Company.

———. [1900] 1988a. *Contending Forces: A Romance Illustrative of Negro Life North and South.* Reprint, in *The Magazine Novels of Pauline Hopkins.* New York: Oxford University Press.

———. [1901–1902] 1988b. *Hagar's Daughter: A Story of Southern Caste Preju-*

dice. Reprint, in *The Magazine Novels of Pauline Hopkins*. New York: Oxford University Press.

———. [1902–1903] 1988c. *Of One Blood: Or, the Hidden Self*. Reprint, in *The Magazine Novels of Pauline Hopkins*. New York: Oxford University Press.

———. [1902] 1988d. *Winona: A Tale of Negro Life in the South and Southwest*. Reprint, in *The Magazine Novels of Pauline Hopkins*. New York: Oxford University Press.

———. [1901] 1991a. "A Dash for Liberty." Reprint, in *Short Fiction by Black Women, 1900–1920*, edited by Elizabeth Ammons. New York: Oxford University Press.

———. [1900] 1991b. "Talma Gordon." Reprint, in *Short Fiction by Black Women, 1900–1920*, edited by Elizabeth Ammons. New York: Oxford University Press.

Johnson, Abby Arthur, and Ronald Maberry Johnson. 1979. *Propaganda and Aesthetics: The Literary Politics of Afro-American Magazines in the Twentieth Century*. Amherst: University of Massachusetts Press.

Kassanoff, Jennie A. 1996. "'Fate Has Linked Us Together': Blood, Gender, and the Politics of Representation in Pauline Hopkins's *Of One Blood*." In *Unruly Voice: Rediscovering Pauline Elizabeth Hopkins*, edited by John Cullen Grueseer. Urbana: University of Illinois Press.

Kevles, Daniel J. 1985. *In the Name of Eugenics: Genetics and the Uses of Human Heredity*. New York: Knopf.

Logan, Rayford W. 1965. *The Betrayal of the Negro: from Rutherford B. Hayes to Woodrow Wilson*. London: Collier-MacMillan.

Murray, Daniel. 1906. "Race Integrity—How to Preserve It in the South." *The Colored American Magazine* 11: 369–77.

Roosevelt, Theodore. 1905. "The American Woman as a Mother." *Ladies' Home Journal* 22 (July): 3–4.

Sundquist, Eric J. 1993. *To Wake the Nations: Race in the Making of American Literature*. Cambridge, Mass.: Harvard University Press.

Tate, Claudia. 1985. "Pauline Hopkins: 'Our Literary Foremother.'" In *Conjuring: Black Women, Fiction, and Literary Tradition*, edited by Marjorie Pryse and Hortense Spillers. Bloomington: Indiana University Press.

Yarborough, Richard. 1988. Introduction. In *Contending Forces: A Romance Illustrative of Negro Life North and South* by Pauline Hopkins. 1900. Reprint, New York: Oxford University Press.

Bad Blood and Lost Borders: Eugenic Ambivalence in Mary Austin's Short Fiction

Penny L. Richards

MARY HUNTER AUSTIN (1868–1934) WAS A PROLIFIC AUTHOR OF STORIES, essays, poetry, drama, and novels, best remembered for her first book, *The Land of Little Rain* (1903). In recent years, Austin has been the focus of renewed scholarly attention, the subject of numerous dissertations, biographies, and reissues (Graulich and Klimasmith 1999; Hoyer 1998; Stout 1998; Ellis 1996). Among the riches being rediscovered, Austin's Western-themed short stories, especially those collected in the anthology *Lost Borders* (1909), have been analyzed through psychological lenses, as autobiographical tales, and as works "enmeshed with the politics of her time"—with postcolonial, feminist, Western, spiritual, and ecological dimensions (Carew-Miller 1999; DeZur 1999, 24–25).

As eugenic themes and language recur throughout these stories, they also lend themselves to consideration in the context of the eugenics movement, with which Mary Austin was indeed "enmeshed," both publicly and privately. She was raised with an acute consciousness of heredity as a public and political topic. Austin's nonfictional writings on the subject were steadfast in their endorsement of popular eugenic policies, especially marriage restriction to prevent unfit unions. But Austin's fiction reflects her personal experience as the mother of a severely disabled child and her struggle with personal questions of guilt and the broader question of her daughter's place in society. As a result, her short stories depict a world in which environment is usually more powerful than heritage; a frontier where even well-meaning laws could have little effect; a space where "mixed blood" and "deformity" are not always unwelcome or tragic. The desert setting of Austin's *Lost Borders* and other stories presents an unforgiving physical climate; the cultural climate, however, includes opportunities to the eugenically "unfit," and challenges to eugenic convictions.

Eugenics in Mary Austin's Life

Mary Austin was raised in an atmosphere charged with eugenic ideas. She was born in the wake of the Civil War in Carlinville, Illinois, to parents active in the temperance movement (Church 1990; Stineman 1989; Fink 1983; Doyle 1939; Austin 1932). After her father's death, her widowed mother took leadership positions in the local Women's Christian Temperance Union. Young Mary accompanied her mother to meetings and gained a wary knowledge of marital relations from the testimony of women wronged by their drunken husbands. By the end of the nineteenth century, the WCTU would develop an effective, widely adopted hygiene pedagogy that stressed the need for care in choosing one's spouse and the harm a flawed father might visit on his unborn children (Cook 1993; Austin 1926). In Mary's day, these same lessons were impressed upon daughters by their Union-trained mothers and communicated in discrete pamphlets.

Other factors in her early life contributed to Austin's emergent eugenic sensibilities. Her mother's brand of Methodism emphasized individual responsibility and traditional sexual morality. Susanna Hunter believed that misfortune was ordained by God as a punishment for individual transgression.[1] Another relevant element of her Illinois upbringing was the presence of triracial seminomadic groups like the Ben Ishmaelites, who still skirted the edges of Midwestern towns in the 1870s, hunting and selling baskets and such. Austin remembered these "mixed breeds" as exotic, stigmatized Others; they were also among the first communities targeted by eugenic programs, beginning in earnest during Mary Austin's teen years (Leaming 1977).[2] Though she had already demonstrated an interest in writing, she chose biology as her major course of studies at Blackburn College in Carlinville; years later, editor Henry Seidel Canby would say of Austin, "She felt that her career, broadly speaking, was in science" (Church 1990, 154–55).

As an adult, Mary Austin remained interested in science, especially the natural sciences; and among them, eugenics held particular fascination. At her death, her library contained the classic works by Lewis Terman, Ellen Key, Margaret Sanger, Eugenio Rignano, and others, as well as books and pamphlets of much narrower distribution.[3] Her writings for newspapers and magazines included articles on subjects associated with eugenics, with titles such as "Talent and Genius in Children" and "Are You a Genius?" (1920s, 1925). Elsewhere, Austin more directly discussed the heritability of traits and the need for eugenic policies. She reviewed books on marriage in the *Birth Control Review* (1927). In a widely noted 1912 address to the women of the New York Legislative League, Austin spoke on "her recipe for happy marriages," including "an investigation of the forbears of each candidate to the marriage, so that data pertaining to mental or physical disabilities could be known and considered in the decision to wed or not to wed" (Pearce 1965, 43). Such pronouncements gained Mary Austin's eugenics-informed positions a great deal of press, with sensational headlines like "Wife Has Right to Boss Home, Says Mary H. Austin," and "Author is Champion of Divorce" (Author unknown 1922, Timmons, und.).

In her 1927 essay "The Forward Turn," Mary Austin explained her advocacy of certain eugenic policies by recalling the dilemmas women faced in previous generations: "Literature of the day make it plain that women were awake to . . . the tragic possibilities involved in bearing children to diseased or dissolute husbands, of continuing a defective strain, the very existence of which was concealed from the young parents under taboos imposed by the terrible innocent ignorance of the Victorian years, and all this before there was any admitted discussion of possible relief" (Austin 1927). Despite her impersonal tone, she might well be remembering her own experience as one of those deceived young parents. Mary Austin's only child was her daughter Ruth (1892–1918), whose severe disability might today be identified as autism (Blend 1996; Hoyer 1995, 246).[4] (Mary placed Ruth in a small private hospital in 1904, where she remained until her death.)

Austin described her motherhood experience publicly, first in an anonymous article and later in her autobiography, in terms that emphasize her own blamelessness and her husband's silent deception: "Brought up as I was, in possession of what passed for eugenic knowledge, it had never occurred to me that the man I had married would be less frank about his own inheritance than I had been about mine. . . . I who had entered motherhood with the highest hopes and intentions had to learn too late that I had borne a child with tainted blood" (Anonymous [Austin] 1927). Assured by doctor friends that she did not cause Ruth's disability, Austin blamed her husband and his family and offered herself as an example of how women are victimized when eugenic protections are not enforced (Doyle 1939, 31; Stineman 1989, 227 n. 43; Austin 1932, 269).

On the basis of the biographical information provided thus far, one might conclude that Mary Austin clearly adhered to the basic eugenic tenets, as understood by an interested layperson in her day: that behaviors, talents, and traits were passed predictably from generation to generation; that biological inheritance far outweighed environment in determining the attainments of the individual; and especially, that restrictive policies should be pursued to improve the human race (Selden 2000; Pernick 1996). But in her short stories, we find a universe where each of these tenets is challenged. Through the medium of fiction, Austin appears to have expressed her doubts, fears, and even hopes about eugenics, disability, and motherhood, even while her nonfictional writings clung to a more confidently eugenic line. These short stories allow, perhaps, a glimpse of the private reservations Austin held about eugenics, reservations developed as a young mother facing her child's disability while living in rural California.

Eugenic Themes in Austin's Short Fiction

Scholars studying Mary Austin's oeuvre often concentrate on her novels, her curious autobiography *Earth Horizon*, or on her genre-defying *The Land of Little Rain*. Her books were certainly popular, but none (at least initially) reached an audience as broad as that of her short stories, which appeared in magazines with circulations in the neighborhood of half a million readers. By 1910, Austin had published more than two dozen short stories, many of which were written when she was a country teacher and struggling parent (Langlois 1990). Although they are at times formulaic, sentimental, and even patronizing, Austin's short stories preserve her early attempts to portray California desert society as a human-environment interface, fraught with pain and possibility. Found throughout the stories are characters, situations, and descriptions that reflect Austin's interest in eugenics. Sickly, disabled, immoral, and mixed-race people populate many of Austin's stories, testing the limits of the eugenic perspective.

The question of whether all traits passed predictably through genetic inheritance haunted Mary Austin as Ruth's mother. As a writer, she passed this burden along to her characters and then created resolutions that absolve parents of guilt. Perhaps the most explicit version of this theme comes from the story "The House of Offence" (Austin 1909d). The unsubtly-named Hard

Mag is a prostitute, whose daughter has lost her respectable guardians and is being returned to the faraway mother she has never known. Hard Mag begs her churchgoing, childless neighbor, Mrs. Henby, to claim the girl instead when she arrives in order to save the child from brothel life (and, she intimates, to save Mag from an encumbrance that would disrupt her own livelihood). Mrs. Henby rejects the idea repeatedly, until Hard Mag pinpoints the neighbor's underlying fear:

> I know what you are thinking, Mrs. Henby. You think there's bad blood, and she will turn out like me maybe, but I tell you it's no such thing. Look here—if it's any satisfaction to you to know—I was good when I had her, and her father was good—only we were young and didn't know any better—we hadn't any feelings except what we'd have had if we had been married—only we didn't happen to—It's the truth, Mrs. Henby, if I die for it. Bad blood! . . . How many a man comes to the House and goes away to raise a family, and not a word is said about bad blood. . . . If it's any question of what she'll come to, you know well enough if I have to take her with me. (Graulich 1987, 88)

Mrs. Henby agrees to take in the little girl, who (she is relieved to discover) does not resemble her mother in the slightest, and the new family thrives.

In this story, Austin personifies her own struggles with motherhood and the question of "bad blood" by splitting attributes of her self between the two main characters (Graulich 1987). Mrs. Henby is given Mary Austin's disappointingly distant marriage and her curiosity about the Other, her domestic expertise, and her maternal longings; while Hard Mag is, like Austin, a socially-suspect working mother who feels certain that she cannot be her daughter's best parent (Hilyer 1993, 99). The daughter is described as being eleven years old, or the age of Ruth Austin in 1903, when Mary Austin began to seek a permanent alternative home for her daughter. As the author of the story, Austin arranged a happy ending for the Henbys, Hard Mag, and her daughter, seeming to confirm Mag's declaration that there's no such thing as "bad blood." However, by explaining the child's conception—by foolish young people, not by sexual deviants—Austin leaves open the possibility that the girl's eugenic credentials are better than they first appear. In that case, Mrs. Henby can believe she is rescuing a truly good child, not suffering a bad woman's tainted offspring. Hard Mag's speech simultaneously ridicules eugenic thinking and submits to its logic. As a mother,

Austin lived this tension, though she found neither the guiltless escape she presented to Hard Mag, nor the fulfilling motherhood she granted to Mrs. Henby.

Mary Austin's desert landscapes are harsh places, where even the basic elements of human identity can be lost or shattered. "To understand the fashion of any life, one must know the land it is lived in," she wrote in "The Basket Maker"(Graulich 1987, 31). Heredity is no match for the environment here among the lost borders. Several of Austin's short stories center on characters who have forgotten themselves—"The Walking Woman," for instance, has forgotten her own name, from its lack of use during her years of wandering (Austin 1907). In a six-part serial,"Blue Moon," a white man, maddened by the desert, forgets that he is white and becomes a Shoshone medicine man (1909b)—the same character is also mentioned in "A Case of Conscience," discussed below (Austin 1909c). A good man becomes a murderer in the desert, but not because the capacity for murder was lurking in his genetic makeup—his body is literally taken over by a murderous spirit, a rather extreme case of environment trumping heredity, found in "The Pocket Hunter's Story" (Austin 1903d).

Austin also portrays desert life as transformative in a positive sense. Far from proposing, as mainstream eugenics did, that environmental reforms could have little or no effect on improving humanity, Austin allows several characters, disabled from birth, to gain skills and flourish beyond ordinary expectations through education and responsible work. "Pahawitz-Na'an" is an orphaned, hemiplegic Paiute boy who is called only "Limpy" by villagers, until his skill as a shepherd saves his people's herds from Shoshone thieves (1903c). "The Little Coyote" is also a shepherd, half-Paiute and "dull of wit," whose single-minded bravery and expert knowledge of his territory save his father's flocks in a snowstorm (1902). Little Coyote sacrifices his life to this goal but in death gains his father's public recognition. A similar story is told of Evaly John, the Paiute main character in "The White Hour" (1903e). Evaly John's parents wish her to become educated among whites, so they send her to the nearby school. Her teacher soon finds that Evaly is "hopelessly dull" (89), but she conceals this finding from Evaly's hopeful parents and protects the girl's feelings by quietly promoting her to remain with her higher-achieving peers. The result is that Evaly never considers herself "dull," but instead shares her father's belief that she is "fit to counsel and instruct" in the secrets of native medicine, one of the "elect and enlightened." This conviction allows Evaly to embark on a desperate journey to save her father's life. As with Little Coyote, the successful exertion costs Evaly's life, but her education has not been wasted. The confidence placed in these characters, by parents, teachers, and community, becomes self-fulfilling.

"Pahawitz-Na'an," "The Little Coyote," and "The White Hour" were all published in the months leading up to Ruth Austin's institutional placement. In all three, Austin appears to be struggling with the question of environment—can a disabled child find a meaningful life if given the opportunity? Evaly, Little Coyote, and Limpy do, but only Limpy (or Pahawitz-Na'an, the adult name he earns) survives the exertion necessary. Years later in 1918, Mary Austin wrote to a close friend about her own daughter's death: "back of her poor, imperfect body, my child's soul waited its deliverance" (Austin Collection). Austin describes her "hopelessly dull" characters, Coyote and Evaly, as having strong bodies—Coyote even "looked a young god . . . with the sun shining on his fine gold-colored limbs" (1902, 252)—but they are also made physically fragile and destined for an early death, as if Austin could not see a long full life for them, or for her daughter. Again, as with the question of "bad blood," the eugenic implications of her stories are ambivalent: environment may indeed have a strong positive impact on some of Austin's disabled characters, but an underlying constitutional weakness remains to defeat them.

In her nonfiction writings, Mary Austin seemed confident in the potential of eugenic policies—marriage restriction, segregation, or sterilization for the unfit—to improve humanity. However, her short stories once again reveal more ambivalence on this point. Certainly, her most autobiographical stories, such as "Frustrate," carry the message that girls should receive more sex education to prevent bad marriages (1912). Elsewhere, however, Austin creates situations in which such policies would be both unworkable and unwelcome. "For law runs within the boundary, not beyond it," she cautions in "The Land," the first story in the *Lost Borders* collection (Graulich 1987, 40). In "The Lost Mine of Fisherman's Peak," she further explains about the setting: "It is so far from the common ways of men by distance, and the manner of living, that nothing disturbs it not native to the soil. Doctrines, schisms, war, politics, the trumpets of reform never reach it" (Graulich 1987, 210–11). Notes Kathryn DeZur, "Austin seeks to demystify imperialist ideology by crossing territorial, biological and narrative 'borders'. . . . She tries to create a kind of borderland where crossover is possible" (DeZur 1999, 22). Some of that crossover involves beings and couplings that violate eugenic recommendations, and generally without dire biological consequences.

One Austin character has a visual disturbance—she sees colorful halos around objects. Whether or not this is a disabling or enabling condition is a central theme of the story "The Coyote-Spirit and the Weaving Woman" (Austin 1904b). Her "infirmity" segregates her from the society of her people, none of whom wishes to live with someone so different. But this outcast has a cheerful, fearless demeanor and a gift for making beautiful baskets, both stemming from her visual disturbance:

> There were some who even said it was a pity, since she was so clever at the craft, that the weaver was not more like other people, and no one thought to suggest that in that case her weaving would be no better than theirs. For all this the basket-maker did not care, sitting always happily at her weaving or wandering far into the desert in search of withes and barks and dyes, where the wild things showed her many a wonder hid from those who have not rainbow fringes to their eyes; and because she was not afraid, she went farther and farther into the silent places until in the course of time she met the Coyote Spirit. (Graulich 1987, 114)

When the Coyote Spirit comes upon the Weaving Woman with murderous intent, the woman literally cannot recognize the spirit as anything but a man, and he is thus rendered powerless to harm her. Eventually, the Weaving Woman convinces the Spirit that he *is* only a man, and he rejoins human society. In this twist on the Red Riding Hood plot, Austin places the question of disability at the center. If a stigmatizing physical difference is actually more a gift than a curse (as it proves to be for the Weaving Woman and for her people), what is the true effect of eugenic practices like segregation?

Austin's California desert is populated by characters identified by their mixed parentage. Marguerita, for example, in "The Bitterness of Women," "was only half-French herself . . . in fact, most people in Tres Piños were a little this or that, with no chance for name-calling" (Graulich 1987, 78). (Note that even her name is a hybrid—neither the French *Marguerite* nor the Spanish *Margarita*, but a combination of these. Austin was a notoriously bad speller, especially of Spanish words, but this particular variation seems intentional.)[5] In several stories, the children of white men and Native women are the objects of custody struggles when the man returns to "civilization." This plot is played for comedy in the short-short "Papago Wedding," for which Austin won an O. Henry award: Susie, the clever mother, insists to the court that none of her five children is the father's offspring, and she won't change her story unless the man (who has a white wife) marries her (Austin 1933).

The same situation receives a more dramatic treatment in "A Case of Conscience" (Austin 1909c). The couple here are a tubercular Englishman, Saunders, "unfit for work or marrying" (Graulich 1987, 48), and a grey-eyed Shoshone woman, Turwhasé, who may be descended from the maddened white medicine man in "Blue Moon"—a pair whose health and heritage would raise several eugenic flags. They form a relationship from which a daughter is born; then Saunders learns that his lung complaint is cured, and he may return to his own society. With his "old, obstinate Anglo-Saxon prejudice that makes a man responsible for his offspring" (48), he leaves with the little girl in tow. As they reenter white civilization, Saunders begins to see the implications of his wilderness idyll: his daughter draws stares, and he wonders if it would be better if the mixed-race "brat" simply died (50). His quandary is resolved when Turwhasé arrives, demanding their daughter's return.

Another approach to this question of mixed parentage is found in "White Wisdom," published late in Austin's life (Austin 1934b). Dan Kearny, a grey-eyed young man raised by a Ute mother and an Irish father, attempts to straddle both worlds in his social life. He is "two-minded" (186), educated in both Ute and white schools, becomes a teacher among his mother's people (men call him Mr. Kearny), and he succeeds in charming both worlds by "two-talking" (188): "He had a White man's way of laughing as he talked, and when the Agent would not agree with him he would say, 'Now, Mr. Malone, as one Irishman to another'; but when the Elders were to be pursuaded he said, 'We Utes.' So he got what he wanted" (188). Even his appearance seems layered. "On his face, on the skin of his body, on his clothes, he was as one painted with Whiteness for the dance of the Sun Returning," the narrator notes (187). Kearny's charm and dual credentials are challenged when he proposes marriage to his long-time love, Miss Bellows, a white woman with rosy skin and turquoise eyes (193). The woman is horrified by his offer, and her father chases Kearny away with the words, "You damned half-breed" (196).

In his shock at this sudden rejection, Kearny rides into the desert, takes on the costume of his Ute friends, marries a Ute woman in a drunken celebration, and declares, "of the blood of Dan Kearny there is little left, Mother, and what is left is the blood of Twice-Bitten, the Gray-Eyed Ute" (197). Only in the story's last paragraph does Kearny's wise aunt confide the truth to her son, the narrator: Dan Kearny never knew that his biological mother was his father's first wife, a white woman, and not the Ute woman who raised him. Austin uses this twist to turn a story about a "tragic hybrid" into a story about the cultural tragedy of racism. Dan Kearny is not destroyed by any internal instability, but by a social practice that restricts marriage to partners whose racial ancestry is assumed to match. (A mixed-race female character who, like Dan Kearny, is educated in various cultural settings is found in Austin's "The Lost Mine of Fisherman's Peak" (1903b), but Lupe is a more straightforward "tragic hybrid," whose racial confusion leads her to murder, madness, and drink.)

Austin questions the wisdom of another eugenic marriage restriction in "Kate Bixby's Queerness," which was only recently published for the first time (Graulich 1987). The title character's mother, Mrs. Bixby, is a suffragist, an officer in the WCTU, and "President of the Society for the Promotion of Eugenics" (278)—in other words, a clubwoman whose sympathies mirror Mary Austin's public positions. Kate Bixby is a teacher, who wishes to marry Francis Whitacre, "a mild, pale brown man" whose "chronic bronchitis" prevents regular employment (277). As Mary Austin was in her early career, Kate Bixby is faced with the widespread preference for hiring only single women as teachers. So, if she marries, she will lose her job, and in any case she cannot marry Mr. Whitacre, because her mother plies him "with pamphlets dealing with heredity"(279) to prevent his asking, thus prompting him to move away. "He was inefficient as a citizen," Austin writes of Whitacre, "and, according to the postulates of Mrs. Bixby's societies, totally unfit for parenthood" (287).

Kate and Francis nonetheless wed quietly, and Kate continues to teach as "Miss Bixby" while hiding her life with Whitacre in San Francisco. The secrecy required to maintain this arrangement gives rise to the "queerness" of the title. She avoids visiting her family, refuses to report on her activities, dresses differently, and otherwise ceases to act like a dutiful single daughter. After wondering about her health and her sanity, the Bixby family finally learns Kate's secret. Her mother's shock and disapproval do not prevent a strong protective impulse from taking hold. When Francis dies several years later, Mrs. Bixby and the rest of the community expect Kate to return to teaching, for which she is again eligible. But Kate makes no effort in that direction, and the reason becomes clear in the last scene: she is expecting a child by her late husband, whose last words to her were, "You will not let them blame me . . . when . . . when they know" (289). The blame he anticipates includes, it is clear, the eugenic disapproval that will greet the news that such a sickly man has fathered Kate's child.

"Kate Bixby's Queerness" was written in 1905, but not published during Austin's lifetime. Her handwritten note attached to the manuscript states that it was "rejected by many editors as too 'radical'" (Graulich 1987, 274). Certainly one objection, noted by Graulich, was the story's implication that married women should not be barred from schoolteaching. The practice of refusing to hire wives was sometimes explained as a eugenic measure—to protect a (presumably) sexually active woman from the physical or mental exertions that might sap an unborn child's vitality.[6] Another disturbing element might well have been the ending, with its unanswered questions about whether Kate will have a healthy baby, and whether the child will be allowed to thrive, in a fretful family that remembers the father's weak constitution. No eugenic consequences befall Kate at story's end; only perpetual uncertainty and the cultural pressures that follow from eugenic thought. Austin leaves readers to ponder the limited power of eugenic policies in the face of human nature, if even good, educated women, like Kate Bixby, are reached and persuaded by eugenic information campaigns, but nonetheless choose "unfit" marriages and conceive children within them.

One of the earliest and most poignant short stories by Mary Austin, "The Castro Baby," presents an alternative community response to disability in a family (1899). In the story, a young mother brings a very sickly infant into a small town for a baby contest. At first, the other women are cruel, and titter over Señora Castro's foolishness; her weak little daughter could never win. Then the women look closer, seeing that the child has been specially dressed to the best of the mother's ability. Their hearts melting, they approach the mother, learn the baby's name, hold her, and admire her clothes. The mothers arrange for the Castro baby to win the contest, and her mother receives the prize (a photo of the little girl). Soon after, the whole town travels to the baby's funeral, grateful in having contributed to the memory of a brief, cherished life. Austin bestows on the Castros the tender, embracing community she and Ruth never knew, a community untroubled by the worst of popular eugenic concerns and suspicions.

What are we to make of the eugenic ambivalence of Mary Austin's short stories? One explanation is found in their setting. For the mainly Eastern magazine audiences, Austin wrote plots and characters that reflected the rural Californian society she knew so intimately. She saw the real, varied social consequences of marriages across racial boundaries.[7] She knew the local disregard for laws made far away, but she also recognized the desire to maintain the appearance of civilized behavior, and she understood that both the disregard and the desire could have dramatic effects. Austin's land of lost borders is not beyond the reach of social pressures, nor beyond the reach of biology, but somehow beyond the norms that strictly Anglo culture imposed. In order to communicate this freer West in her short stories, Austin drew on the real complexities she saw, even (or perhaps especially) when they challenged her eugenic sensibilities.

Another source of the ambivalence found in Austin's fiction is surely her motherhood experience. As Ruth's mother, Mary Austin could not be dispassionate or simplistic on the subject of inheritance, disability, and public policy. While her later public persona was consistently pro-eugenic, her earlier short stories betray an impulse to hope for other explanations for her daughter's status, other resolutions for her own pain and guilt. For credibility's sake, her essays had to be consistent, but fiction imposed no such restriction. Through short fiction, Austin the scientist could experiment with her characters' lives; her own life resisted such easy tinkering. She played the same situation over and over, each time with a variation in the basic ingredients and outcome. So there are multiple stories of white men and native women having children, multiple stories in which the main character is disabled and stigmatized, and multiple stories where mothers and children become separated by fate. We can read them as Mary Austin's carefully coded laboratory notes, as she sought an elusive formula for personal redemption from her sorrowful motherhood, an alternative to the eugenic beliefs that offered little solace.

In the short stories of Mary Austin, the literary expression of eugenic thought did not emerge solely from her education, her reading interests, or casual contacts with theory. The stories are part of a whole life, full of contradictions, false starts, disappointments, and successes. Her story adds a dimension to the history of eugenics, as a movement and as a literary theme, by illustrating the deeply personal impact of the scientific and social rhetorics of eugenics. Because her experiences with "tainted inheritance" were firsthand and painful, Austin's stories about family, inheritance, and disability, even at their most formulaic, brim with ambivalence and complexity.

NOTES

The research reported in this paper was accomplished in large part during my appointment as an Andrew W. Mellon Foundation (I) Fellow at the Huntington Library, San Marino, California. The Center for the Study of Women at the University of California at Los Angeles has generously granted me Research Scholar status for the pursuit of this project, as well. Emily Abel, Dona Avery, Judy Singer, and others have provided helpful comments during the preparation of this essay.

1. The harsh effect of this belief is illustrated in her response to disability: when she learned that Mary Austin's child was disabled, Susanna Hunter wrote to her distraught daughter, "I don't know what you've done, daughter, to have such a judgment upon you." They never spoke again (Austin 1932, 257).

2. Austin records her memory of these groups' presence in her notes on Illinois history, in preparation for her autobiography: "Few free negroes with Indian blood and other mixed breeds still found in the vicinity of C[arlinville], making baskets and living off the land" (Austin Collection; the first folder of box 122, notes on the years 1884 to 1885).

3. The following books, among others, are listed in an "Inventory of House and Library" (Austin Collection, box 122): Kellogg, *Darwinism Today*, Nystrom, *The Natural Laws of Sexual Life*, Terman, *The Measurement of Intelligence* and *Genetic Studies of Genius*, Rignano, *Upon the Inheritance of Acquired Characters*, Holmes, *The Trend of the Race*, Dell, *The Outline of Marriage*, Blacker, *Birth Control and the State*, Key, *Love and Ethics* and *Love and Marriage*, and Sanger, *Women and the New Race*.

4. Ruth's characteristics, as described firsthand by Doyle (1939), pp. 167–72, include strange vocal sounds, prolonged crying and screaming without known cause, hard-to-manage behaviors, and a failure to develop self-care skills. A reference by Doyle to "restless uncertain movements of her small hands" might suggest Rett syndrome as another possible diagnosis.

5. An interesting phenomenon in Austin's correspondence is the appearance of several letters from readers offering to proofread Austin's future works, for example, Mary A. Bucknam to M. Austin, 6 March 1933 (Austin Collection).

6. The idea that a pregnant woman's thoughts and actions can influence the development of her unborn child is ancient; the idea that the state should protect the child from its mother's mental or physical overexertion as a eugenic measure was a nineteenth-century development (Rosenberg 1974). Austin's autobiography includes a lengthy discussion of this notion of "the preciousness of women" (Austin 1932, 145–46).

7. For example, acquaintance Peter Watterson became involved with a part-Indian woman, and he eventually committed suicide as a result of the social pressures, wrote his mother in a letter to Mary Austin, 2 June 1907 (Austin Collection).

References

NOTE: In the case of Austin's short stories, the original published appearance is cited, where possible; appearances in anthologies are noted in brackets afterwards, using the following abbreviations:

BB	*Beyond Borders* (Ellis 1996)
BW	*Basket Woman* (Austin 1904)
LB	*Lost Borders* (Austin 1909)
LOLR	*Land of Little Rain* (Austin 1903)
MAR	*Mary Austin Reader* (Lanigan 1996)
MOF	*Mother of Felipe and Other Early Stories* (Austin 1950)
OSS	*One-Smoke Stories* (Austin 1934)
SCLB	*Stories from the Country of Lost Borders* (Austin 1987)
WT	*Western Trails* (Graulich 1987)

All the anthologies are themselves included in this list of references.

Austin, Mary. 1892. "The Mother of Felipe." *Overland Monthly* 20 (November): 534–38. Full text available online at http://etext.lib.virginia.edu. [MAR, MOF]

———. 1899. "The Castro Baby." *Black Cat* 42 (March): 27–31. [WT]

———. 1902. "The Little Coyote." *Atlantic Monthly* 89 (February): 249–54. Full text available online at http://etext.lib.virginia.edu.

———. 1903a. "The Basket Maker." *Atlantic Monthly* 91 (February): 235–38. Full text available online at http://etext.lib.virginia.edu. [LOLR, MAR, SCLB, WT]

———. 1903b. "The Lost Mine of Fisherman's Creek." *Out West* 19 (November): 501–510. [MOF, WT]

———. 1903c. "Pahawitz-Na'an." *Out West* 18 (March): 337–40.

———. 1903d. "The Pocket Hunter's Story." [LB, SCLB]

———. 1903e. "The White Hour." *Munsey's Magazine* 29 (April): 88–92. Full text available online at http://etext.lib.virginia.edu.

———. 1904a. *The Basket Woman*. Boston and New York: Houghton Mifflin.

———. 1904b. "The Coyote Spirit and the Weaving Woman." [BW, MAR, WT]

———. 1907. "The Walking Woman." *Atlantic Monthly* 100 (August): 216–20. Full text available online at http://etext.lib.virginia.edu. [LB, SCLB, WT]

———. 1909a. "The Bitterness of Women." *Harper's Weekly* 53 (October 9): 22–23. Full text available online at http://www.angelfire.com/wa2/buildingcathedrals/. [LB, SCLB, WT]

———. 1909b. "The Blue Moon." *Sunset Magazine* 21 (January–June): pages unknown. (A serial in six parts.) Clippings found in Austin Collection, Huntington Library (box 124).

———. 1909c. "A Case of Conscience." *Harper's Weekly* 53 (14 August): 22–23. [LB, SCLB, WT, MAR]

———. 1909d. "The House of Offence." *Harper's Weekly* 53 (23 October): 22–23. [LB, SCLB, WT]

———. 1909e. "The Land." [LB, SCLB, WT]

———. 1909f. *Lost Borders*. New York and London: Harper and Brothers.

———. 1912. "Frustrate." *Century Magazine* 83 (January): 467–71. Full text available online at http://etext.lib.virginia.edu. [WT, MAR]

———. c. 1920s. "Talent and Genius in Children." *Children, the Magazine for Parents*. Exact date and pages unknown. Clipping in the Austin Collection, Huntington Library (box 128, folder labeled "Genius").

———. 1925. "Are You a Genius?" (Ruth Laughlin Barker interviews Mary Austin). *Success* (November). Pages unknown. Clipping in the Austin Collection, Huntington Library (box 128, folder labeled "Genius").

———. 1926. "Amorousness and Alcohol." *Nation* 122 (23 June): 691–92.

———. 1927. "The Forward Turn." *Nation* 125 (20 July): 57–59.

———. 1927. Review of Keyserling, *The Book of Marriage. Birth Control Review* (May). Pages unknown. Copy found in Austin Collection, Huntington Library (box 129, folder labeled "Marriage and Birth Control and Women").

———. Anonymous [Mary Austin]. 1927. "Woman Alone." *Nation* 124 (2 March): 228–30. [MAR, BB]

———. 1932. *Earth Horizon*. New York and Boston: Houghton.

———. 1933. "Papago Wedding." *Golden Book* 17 (March): 262–64. [OSS, WT]

———. 1934a. *One-Smoke Stories*. Boston and New York: Houghton Mifflin.

———. 1934b. "White Wisdom." [OSS, WT]

———. 1950. *The Mother of Felipe and Other Early Stories*. Edited by Franklin Walker. San Francisco: Book Club of California.

———. 1987. "Kate Bixby's Queerness." [WT]

———. 1987. *Stories from the Country of Lost Borders*. Reprint, with an introduction by Marjorie Pryse. New Brunswick, N.J.: Rutgers University Press.

———. Manuscripts, clippings, and correspondence. Austin Collection, Huntington Library, San Marino, California.

[Author unknown]. 1922. "Wife Has Right to Boss Home, Says Mary H. Austin; Noted Novelist Thinks Sacrifices Earn this Reward." *American Pictorial* (November 3). Clipping found in Austin Collection (box 126, folder labeled "Personal Interviews").

Blend, Benay. 1996. "Building a 'House of Earth': Mary Austin, Environmental Activist and Writer." *Critical Matrix* 10: 73–89.

Carew-Miller, Anna. 1999. "Between Worlds, Crossing Borders: Mary Austin, Liminality, and the Dilemma of Women's Creativity." In *Exploring Lost Borders: Critical Essays on Mary Austin*, edited by Melody Graulich and Elizabeth Klimasmith. Reno: University of Nevada Press.

Church, Peggy Pond. 1990. *Wind's Trail: The Early Life of Mary Austin*. Santa Fe: Museum of New Mexico Press.

Cook, S. A. 1993. "Educating for Temperance: The Women's Christian Temperance Union and Ontario Children, 1880–1916." *Historical Studies in Education* 5: 251–77.

DeZur, Kathryn. 1999. "Approaching the Imperialist Mirage: Mary Austin's 'Lost Borders.'" In *Exploring Lost Borders: Critical Essays on Mary Austin*, edited by Melody Graulich and Elizabeth Klimasmith. Reno: University of Nevada Press.

Doyle, Helen M. 1939. *Mary Austin: Woman of Genius*. New York: Gotham House.

Ellis, Reuben J., ed. 1996. *Beyond Borders: The Selected Essays of Mary Austin*. Carbondale: Southern Illinois University Press.

Fink, Augusta. 1983. *I-Mary: A Biography of Mary Austin.* Tucson: University of Arizona Press.

Graulich, Melody, ed. 1987. *Western Trails: A Collection of Short Stories by Mary Austin.* Reno: University of Nevada Press.

Graulich, Melody, and Elizabeth Klimasmith, eds. 1999. *Exploring Lost Borders: Critical Essays on Mary Austin.* Reno: University of Nevada Press.

Hilyer, Barbara. 1993. *Feminism and Disability.* Norman: University of Oklahoma Press.

Hoyer, Mark T. 1995. "Prophecy in a New West: Mary Austin and the Ghost Dance Religion." *Western American Literature* 30: 237–57.

———. 1998. *Dancing Ghosts: Native American and Christian Syncretism in Mary Austin's Work.* Reno: University of Nevada Press.

Langlois, Karen S. 1990. "A Fresh Voice from the West: Mary Austin, California, and the American Literary Magazines, 1892–1910." *California History* 69 (Spring): 22–35.

Lanigan, Esther F., ed. 1996. *A Mary Austin Reader.* Tucson: University of Arizona Press.

Leaming, Hugo P. 1977. "The Ben Ishmael Tribe: A Fugitive 'Nation' of the Old Northwest." In *The Ethnic Frontier: Essays in the History of Group Survival in Chicago and the Midwest,* edited by Melvin G. Holli and Peter d'A. Jones. Grand Rapids, Mich.: William B. Eerdmans Publishing.

Pearce, Thomas M. 1965. *Mary Hunter Austin.* New York: Twayne Publishers.

Pernick, Martin. 1996. *The Black Stork: Eugenics and the Death of "Defective" Babies in American Medicine and Motion Pictures since 1915.* New York: Oxford University Press.

Rosenberg, Charles. 1974. "The Bitter Fruit: Heredity, Disease, and Social Thought in Nineteenth-Century America." *Perspectives in American History* 8: 189–235.

Selden, Steven. 2000. "Eugenics and the Social Construction of Merit, Race, and Disability." *Journal of Curriculum Studies* 32: 235–52.

Stineman, Esther Lanigan. 1989. *Mary Austin: Song of a Maverick.* New Haven, Conn.: Yale University Press.

Stout, Janis P. 1998. *Through the Window, Out the Door: Women's Narratives of Departure, from Austin and Cather to Tyler, Morrison, and Didion.* Tuscaloosa: University of Alabama Press.

"Feeble-Minded" White Women and the Spectre of Proliferating Perversity in American Eugenics Narratives

Elizabeth Yukins

> Comparatively few persons yet realize the suffering, the moral degredation [sic], and not least, the increasing expense entailed upon the public by the progeny, often illegitimate, of feeble-minded women. Could our citizens know the truth, the enormous expense, and the depth of degredation caused by this group of degenerates, they would be amazed. Could they look into the future and see what would be the accumulated cost piled up before them in money, in immorality, in succeeding generations of defectives, they would not rest until they had sought means to prevent all this.
>
> —Amos Butler, Secretary of the Board of State Charities in Indiana

IN AN ADDRESS TO THE AMERICAN ASSOCIATION FOR THE ADVANCEMENT of Science, published in 1901, Amos Butler offered a dire warning about the danger of a certain type of fertile woman: "One perverted feeble-minded woman can spread throughout a community an immoral pestilence which will affect the homes of all classes, even the most intelligent and refined."[1] Butler's ominous claim echoes the sentiment shared by many prominent eugenic public policy advocates at the turn of the century—reproduction was a crucial site for gender, class, and race regulation, and national well-being depended upon careful genealogical surveillance. As the source of reproduction, a woman's body promised either the continuing progression of genealogical, and thus national, development, or the insidious threat of moral pathology and biological degeneration. What Butler's eugenic warnings make clear is that women's bodies were believed to carry invisible genetic traits, and such dissonance between visible appearance and underlying characteristics caused significant class and race anxieties even beyond traditional black and white racial categorizations and fears. Butler's vocabulary of contact and contagion points to the imagined connection between moral pestilence and biological degeneration, and his provocative imagery comes out of his, and his constituents', work on controlling the lives and bodies of impoverished, often unmarried, white mothers.

By the end of the nineteenth century, politicians and social scientists had begun to define single mothers and their progeny not simply as transgressors of societal moral codes, but as distinctly different in their biological composition. Women who had children out of wedlock were believed to carry a flawed blood type, and their children were perceived as both evidence of biological degeneration and future perpetuators of hereditarian weakness.[2] Despite the fact that late nineteenth- and early twentieth-century public debates about economic aid to single mothers focused on the experiences of white women and essentially excluded black women from all access to economic relief (Gordon 1994), the concept of "whiteness" that justified state aid and legitimated maternal need was neither homogeneous nor equitable. Within the generic social category of "white women" was a highly fractured class system that differently evaluated the racial worth of middle- and upper-class white women and impoverished white women. This essay examines how unmarried mothers of the latter class were pathologized as dangerous biological contaminants.

The focus on whiteness in this essay is meant neither to elide nor to detract from the political discrimination and social injustices perpetrated against black women in the early decades of this century. In dominant discourses about sexuality, morality, and motherhood, black women were vilified and denied rights by a representational system that justified oppression under the auspices of objective evaluation and necessary control. In debates about "women," gender similarities between black and white women were mitigated by white presumptions about inherent racial difference. It is the question of difference that I seek to explore in this essay, but I intend to invert the strictures of white surveillance to examine whiteness's own cracks and fissures: to deconstruct white claims to homogeneity and to reveal how intraracial class and gender differences were medicalized and rendered suspect. By exploring the genealogical myths developed in two influential family studies, I demonstrate how race and class biases converged in scientific and popular literature to designate the sexual agency and reproductive powers of poor white women as significant threats to Anglo-American racial purity and national sovereignty.

The texts I examine to illustrate this imagined threat are a late nineteenth-century anthropological study of a poor family in rural New York known as the Jukes, and an early twentieth-century eugenics narrative about the history of a family in New Jersey named the Kallikaks. Two similar themes that arise in these texts that indicate important information about eugenic strategies in America are the authors' efforts to identify genealogical origins and to diagnose reproductive tendencies. While such efforts reveal eugenicists' investigative interests in diagnosis and containment—making visible previously unknown genetic traits and regulating their transmission—the hereditarian theories developed in these two texts also indicate the very instability of such a prescriptive endeavor. In order to maintain supposed white racial superiority, eugenicists had, in fascinating ways, to protect white blood from itself. To illustrate the paradox of this social agenda, I employ two phrases, "intraracial miscegenation" (white class mixing imagined as biological alteration) and "proliferating perversity" (the growth of non-normative, transgressive reproductive activity), to examine the perceived dangers of class proximity, racial similitude, and unregulated reproduction in eugenicists' genealogical imaginary. I argue that the convergence of class difference and women's reproductive agency must be understood as a central nexus in eugenic anxieties about social change, for in the influential stories about the Jukes and the Kallikaks it is lower-class women's reproductive bodies that are identified as dangerously white.

Bad Blood and Juke Genealogy

Single mothers held a central position in turn-of-the-century scientific and political debates about social ills and their reform. These mothers and their children were highly analyzed subjects of sociological, anthropological, and biological research: Who were the women who became single mothers? What were the causes of their downfall? What influence did they have on the moral fabric of the community, and, in turn, how did their illegitimate children affect the cultural and economic future of the nation? Beginning in the nineteenth century in Britain and quickly growing in popularity in America, eugenic scientific theory gained the forefront in debates about both single motherhood and social nonconformity in general. By using value-laden assessments of "blood types," racial ancestry, and estimated mental capacities, eugenicists constructed biological explanations for complex social issues. In terms of single mothers, eugenicists argued that moral transgression was attributed to inherited mental incapacity, and that such mothers would, in turn, pass their flawed blood onto their illegitimate children. Eugenicists' warnings about degenerative reproduction reached a large audience and gained significant popularity by means of social welfare literature, book-length studies, newspaper reports, and mainstream magazine articles. At stake, they argued, was the purity and sanctity of the nation's blood stream—single mothers and their children had to be controlled before further moral, biological, and national deterioration occurred.[3]

The premise that like produces like shaped American eugenic philosophy and fueled an expanse of research, theorization, and publication. Although social scientists and biologists had categorized and alienated persons of non-northern European heritage (such as African Americans, eastern Europeans, and Jews) as inherently racially different, they still faced the problem of social defect among those persons they defined as genetically superior, i.e., white Anglo-Saxon Protestant Americans.[4] What did it mean when white persons of American lineage and northern European ancestry did not attain or retain economic and cultural authority? Social reformers who sought an explanation for poverty and other forms of "human failure" were troubled by the genetic implications of poor whites, white prostitutes, white alcoholics, white criminals, and the like. The assurity of white racial superiority, a myth constructed during America's socioeconomic development as a slaveholding nation and further edified by racist social-science theories, was confronted at the end of the nineteenth century with a growing (i.e., increasingly visible) problem of economic impoverishment and human misery. Dominant whiteness was threatened with the spectre of inadequacy and fallibility within its own racial borders. In order to account for this difference and still maintain the ideology of white superiority, eugenicists developed a hereditarian explanation based on individual genetic weakness rather than racial inferiority: "degenerate" whites threatened the purity of superior white "germ plasm," but they signified genetic flaws rather than a racial type.[5]

The concept of hereditary social failure, which eventually developed into the genetic theory of feeblemindedness, first became widely known through Richard Dugdale's much-publicized

1877 report, *"The Jukes," A Study in Crime, Pauperism, Disease, and Heredity* (1975). In the mid-nineteenth century, Dugdale studied the inhabitants of a rural New York jail system and, in the role of criminal anthropologist, researched the genealogical history of a poor, white family living in a mountain region of western New York state whom he dubbed "the Jukes." In his report he describes in bleak language the story of a family of 709 persons who purportedly birthed, over the course of six generations, 18 brothel-keepers, 76 convicted criminals, 128 prostitutes, and over 200 relief recipients (1975, 68). Given these statistics, he estimates that the total cost to the public due to welfare relief, incarceration procedures, medical aid, and other expenses was $1,308,000: "Over a million and a quarter dollars of loss in 75 years, caused by a single family 1,200 strong, without reckoning the cash paid for whiskey, or taking into account the entailment of pauperism and crime of the survivors in succeeding generations, and the incurable disease, idiocy and insanity growing out of this debauchery, and reaching further than we can calculate" (70).

While Dugdale himself was neither a eugenicist nor a strict hereditarian (he believed that human characteristics resulted from the influences of both heredity and environment), his analyses of the Juke family provided welcome fodder for the eugenicists who were rapidly gaining social and political power. For the eugenicists, the message to be derived from Dugdale's study was clear: if the first Jukes had been prevented from breeding, the state and its taxpayers would have been spared a large and costly burden. Links between Dugdale's work and the eugenicists' use of his findings were made clear in a follow-up study, *The Jukes in 1915* (1916), published by two leading eugenicists, Arthur Estabrook and Charles Davenport, in which the authors built upon Dugdale's "Jukes" research in order to advocate for eugenic segregation and sterilization of "feeble-minded" persons (Estabrook 1916). Dugdale's initial study proved useful for the eugenicists because, although Dugdale recognized poverty as having circumstantial causes, he also diagnosed pauperism as a form of inherited disease. In his report he imagines the following cause-and-effect continuum: "Hereditary pauperism rests chiefly upon disease in some form, tends to terminate in extinction, and may be called the sociological aspect of physical degeneration. . . . the debility and diseases which enter most largely in its production are the result of sexual licentiousness. . . . Induced [circumstantial] pauperism may lead to the establishment of the hereditary form" (Dugdale 1975, 38). Dugdale thus offers an image of personal weakness generating social disease, and, even more significantly, the social contaminant is located in the blood. The common denominator linking the people in Dugdale's essay is "Juke blood," and Dugdale describes how this blood mixed widely throughout the community by means of "harlotry," "prostitution, with its complement of bastardy," sexual crimes, and ill-chosen marriages (13).

The instability of blood and its potential to transgress imagined genealogical boundaries linked Dugdale's narrative of the Juke family with eugenicists' theories of degeneration. For Dugdale's contemporaries, signs of social degeneration indicated reproduction-gone-wrong. Because many Victorian scientists either misread Darwin's theory of evolution (Darwin rejected any de facto relation between evolution and progress, and instead preferred the description "descent with modification") or combined it with Herbert Spencer's theory of "survival of the fittest," they often "equated organic change with organic progress" (Gould 1977, 36–37). Eugenicists in particular envisioned a biological ideal in which selective human breeding could produce an ever-stronger, ever-more advanced white race. Hereditary degeneration, with its connotations of flawed blood and backwards movement, symbolized not only racialized genetic failure within this evolutionary schema, but also the potential for a continued falling-off due to the degenerative powers of moral contamination. The *Oxford English Dictionary* cites both a scientific and a general definition of "degeneration," and, when read alongside one another, the two definitions offer insight into how evolutionary hierarchies are structured by progressionist judgments. The biological sciences define degeneration as "A change of structure by which an organism, or some particular organ, becomes less elaborately developed and assumes the form of a lower type"; the general definition of degeneration reads, "the falling from ancestral or earlier excellence; declining to a lower or worse stage of being; *degradation of nature*."[6] The latter definition, which construes biological progress as a hierarchized telos, is particularly significant in terms of how degeneration is presented not merely as an evolutionary failure, but also as a contradiction, an affront, or more aptly, a perversion of nature.

My argument linking degeneration to social anxieties about perverse (re)productivity owes much to Jonathan Dollimore's provocative analysis of the perverse as "aberrant movement." In *Sexual Dissidence* (1991), Dollimore argues that social condemnations of perversity illustrate how a subject is labeled perverse when said subject is deemed to have strayed and deviated from the dominant order. He identifies this condemned movement as the "paradoxical perverse," for the subject identified and alienated as perverse had to have moved away from, and thus developed out of, conditions otherwise defined as normal (1991, 103, 106–7, 116). The label of perversion thus paradoxically serves both to alienate suspect identities and also to create a visible, vital connection between "normal" development and its "perverse" alteration. In terms of scientific theories about sexuality and family structure, the moral connotations elicited by the definition of degeneration as a "degradation of nature" help to explain turn-of-the-century eugenicists' attitudes toward reproduction. There was, for these scientists, little or no distinction between reproduction and morality, and they promoted a theory of nature and natural selection in the interest of "race improvement" as a higher moral good. Instances or types of human breeding that contradicted eugenic principles of selection or produced "dysgenic" progeny (i.e., persons deemed genetically weak) were considered acts of sabotage against "the race," and, consequently, a profoundly debilitating crime against nature. As a prototype for later studies of "inherited degeneracy," Dugdale's report on the Jukes and his characterizations of feeblemindedness offered much that seemed "unnatural" to middle-class American social norms. The reported actions of the Juke family included a range of social crimes, the foremost of which was unregulated "sexual excess" and reproduction.[7] In a general survey of Juke family characteristics, Dugdale states that "fornication, either consanguineous or not, is the backbone of their habits"(1975, 13). He then goes on to link their sexual proclivities to pauperism, crime, prostitution, intemperance, and disease. For Dugdale, the compulsion to fornicate is both cause and effect of the Jukes' degenerative tendencies: sexual excess creates a "profligate" and "contaminating" environment, and, within such an immoral setting, persons substitute the pursuit of "licentiousness" (60) for the rewards of industrious labor. Dugdale's ideological biases and his rhetorical techniques become even more apparent when he summarily denounces the Jukes' homes as "log huts and hovels which . . . form hot-beds where human maggots are spawned" (61).

With this imagery of nests and maggots, Dugdale depicts a form of perverse domesticity, a domesticity that expands and proliferates beyond the bounds of the "governed hearth."[8] In this scene, family members indiscriminately mate, community inhabitants "spawn" (rather than reproduce), and the resultant progeny represent social parasites who feed upon the state's misguided attempts at civic reform. Such a familial configuration would not only appear deviant and abhorrent to Dugdale's nineteenth-century middle- and upper-class reading audience, but also deeply threatening to their sense of cultural security and social boundaries. Contrary to Dugdale's personal conjecture that the Juke family would degenerate into extinction, his report instead describes expanding lines of family descent that seem to warn of prodigious reproduction and proliferating perversity. This image of unbounded, generative immorality conflicted sharply with what Gwendolyn Mink has termed the nineteenth-century American "canons of republicanism and domesticity" (1990, 97). These two interdependent ideologies construed nineteenth-century American political life as a masculine affair and positioned women in the home as the (re)producers and caretakers of the nation's young citizens. Mink argues that American republicanism generated racialized ideals of American citizenship that placed white men in control of the democratic state and situated white womanhood as the "guardian of male virtue and reproducer of (white) republican order" (93). This ideological system thus secured a dualistic public/private paradigm that relegated women's experiences and responsibilities to the privatized sphere of domesticity, but narrowly allowed white women to participate in the political community via reproduction and "socialized motherhood" (Mink 1990, 97).

The reproductive practices of the Juke community, however, seriously disrupted tidy, racialized ideals of republican domestic organization. I will examine the role of women in Dugdale's narrative in a moment, but first it is important to recognize how *the whiteness* reproduced in the Jukes' homes represented a perverse reconfiguration of the ideal of racial homogeneity pursued by Anglo-American nativists. Dugdale describes the family's homogeneity and distinctly American lineage as follows: the first six Jukes he encountered were in the New York prison system—"These six persons belonged to a long lineage, reaching back to the early colonists, and had intermarried so slightly with the emigrant population of the old world that they may be called *a strictly American family*"[9] (1975, 8; my emphasis). While this "strictly American family" might have debunked middle-class presumptions about colonial lineage and racial superiority, the Jukes example was instead rendered pathological by means of a biological theory. The type of whiteness promulgated by the Jukes was defined by eugenicists as a "weak stock," a strain of "bad germ-plasm" that rendered its carriers "not fit . . . for participation in a highly organized society."[10] Because Dugdale identified the first Juke in the family genealogy as "a descendant of early Dutch settlers" (1975, 14), later eugenicists could not disassociate the "weak stock" from its northern European origins. Given this limitation on geographic scapegoating, the resultant eugenic strategy was instead to further hierarchize racial categories by subdividing white blood into differentiated germ plasm. Possessors of good white blood—persons who achieved or maintained social and economic success—needed to band together to contain and eradicate carriers of bad white blood (the latter blood type could be identified by the human form it produced: persons in need of charity relief, in the state criminal system, etc.). Yet, while this eugenic schema accounted for and distanced the "social failure" of those persons who would be judged "white" in the visual field, it also exposed whiteness as an unstable racial category: not only could some unseen substance embedded within whiteness prove degenerative, but, if whiteness can be fractured and hierarchized into differing biological types, then the threat of white-white miscegenation (otherwise known as class mixing) looms large.

In one of the report's first sections on Juke "traits," Dugdale declares that "[t]he distinctive tendency of the Juke family is . . . harlotry and bastardy" (1975, 18). While Dugdale never explicitly states that women are responsible for heredity degeneration, he signals maternal fault in his discussion of "harlotry"—a general label for female sexual immorality—and its degenerative effects. He writes that "Harlotry may become a hereditary characteristic and be perpetuated without any specially favoring environment to call it into activity" (26), and he offers a number of statistical tables that exhibit statistics on the amount of "Harlotry in the 'Juke' blood" (18–19). The use of statistics is important in Dugdale's project, for, according to his calculations, harlotry is "over twenty-nine times more frequent with the Juke women than in the average of the community" (18–19). (Dugdale's calculations are based upon inference, speculation, and comparison with "the community"; what "community" he uses as his point of comparison is unclear.) A crucial degenerative effect of harlotry, Dugdale stresses, is the production of illegitimate children. In his teleology of harlotry, bastardy, and degeneration, Dugdale argues that female "licentiousness" produces illegitimate children and illegitimacy is an important causative factor in crime. A synthesis of Dugdale's "inductions" might read: not all bastards are criminals, but many criminals are bastards.[11] A woman's harlotry thus spawns a continuum of social transgressions: the begetting of illegitimate children represents a moral offense (bastardy) that corporealizes a biological offense (degeneration) that leads to a cultural offense (crime). According to Dugdale's reasoning, bastardy is thus the paradoxical condition of generative degeneration: while bastards mark female sexual transgression, these stigmatized signifiers are themselves able to reproduce (Dugdale pays particular attention to the illegitimate daughters of unmarried women as he seeks to ascertain their tendency to bear further illegitimate progeny). If bastardy represents a continuum of reproduction-gone-wrong, in the sense that the illegitimate progeny repeat the reproductive transgression, then to trace backwards the effects of the effects is to locate "original sin" in female sexuality and what can be conceptualized as perverse motherhood, i.e., a type of motherhood that spawns cultural deviation. In Dugdale's study, the genealogical search backwards concludes in an original mother-figure, Ada Juke, who becomes the fatalistic symbol for and cause of sexual transgression, domestic failure, and the perverted lineage of her progeny.

While Dugdale establishes that the Juke family history began with Max, the early Dutch settler, he begins his genealogical study of the family's heredity patterns with the progeny of Max's five daughters, Ada, Bell, Clara, Delia, and Effie (Dugdale gives pseudonyms to the daughters so that their birth order corresponds to the first five letters of the alphabet). Although Dugdale admits that the early genealogical history of Max Juke is vague and virtually impossible to ascertain, he states that Max had at least two sons and numerous other children of legitimate and illegitimate status. Despite these genealogical uncertainties, Dugdale nevertheless designates the five daughters as the original sources for what comes to be known as "Juke blood" and traces Juke maternal lineage through five generations. The attribution of reproductive responsibility is exemplified in Dugdale's description of Ada Juke's lineage. Dugdale reports that Ada Juke is "better known to the public as 'Margaret, the mother of criminals'" (1975, 15). Ada is named as such (by both "the public" and what seems to be a silent concurrence by Dugdale) because she gives birth to an illegitimate son who, in Dugdale's words, "is the progenitor of the distinctively criminal line" (15). The first two large fold-out genealogical charts included in Dugdale's report schematize the lineage of Ada Juke. Both charts situate Ada Juke as the originating, primary source of Juke lineage (Max Juke is not even listed on the charts), and the first chart maps out seven generations of "The Illegitimate Posterity of Ada Juke." As the mother of a bastard child, some of whose progeny became criminals and the parents of illegitimate children themselves, the figure of Ada Juke serves as the ideological intersection of moral injunction and biological theory: her reproductive deviance becomes constructed as an originary site of genetic degeneration. In other words, Ada Juke's sexuality becomes a powerful symbol for a double-edged crime against nature, and her bastard child becomes both a marker of past transgression and a signifier of future crime.

Goddard's Myth

These themes of transgressive female sexuality, perverse motherhood, and criminal genealogy were expanded upon and exploited by later eugenicists. The lessons taught by the Juke narrative were supplemented with other genealogical studies, and theories of degeneration evolved into scientific investigations of *feeblemindedness*. This term, first coined in England and made popular in America by the prominent eugenicist, Charles Davenport, was used as a diagnosis for suspect mental capacities and a wide range of socially deviant behaviors. The feebleminded were those people who did not, or could not, make a proper social fit; something in their genetic makeup (and this reason remained vague) caused them to fail in the economic and social opportunities provided by America's democratic-capitalist system. One of the foremost authorities on feeblemindedness and its heredity patterns was Henry Goddard, the director of the research laboratory at the Training School for Feeble-Minded Girls and Boys in Vineland, New Jersey. Goddard was among a group of biological, medical, sociological, and anthropological scientists who revised the focus of modern American eugenics. In the early part of the twentieth century, the eugenics movement altered its agenda from the desire to encourage breeding in families who were deemed socially superior to a desire to halt reproduction in those families who were identified as inferior and degenerative.[12] Under the latter category, solutions to "unworthy" reproduction included mandating state records on mental capacities and moral traits, strict prohibitions on marriage, segregation, sterilization, and restrictions on immigration.

In a 1912 article in the social work magazine, *The Survey*, Goddard estimated that the United States had an "army of over five hundred thousand . . . social misfits" who could be classified as feebleminded (1912, 1852). He warned that these persons were an insidious and highly dangerous threat to the (white) race and nation, for they not only continually increased their "army" through prodigious reproduction, but, as white citizens, they also bred *within* the racial and national borders. In other words, they could not be kept out, for they were already inside. Goddard warned, "Our government spends hundreds of thousands of dollars examining immigrants to see that none who are suffering from contagious diseases or who are paupers or feeble-minded are admitted to the country; but here is a group already within our country who are breeding a race of feeble-minded people more dangerous than many barred by the immigration inspectors" (1912, 1853). Goddard's and his constituents' remedy was to create new, internal boundaries between citizens, and the identification and segregation of "feeble-minded" persons became the imagined means by which to purify and reify national identity.

Goddard attained both personal fame and widespread public concern about feeblemindedness with his 1912 study entitled *The Kallikak Family* (1973).[13] In this narrative of one family's genealogical history, Goddard utilized Dugdale's techniques of historical reconstruction, anthropological supposition, and scientific storytelling to portray a family whose notoriety would come to supplant the Jukes' position in America's social imaginary.[14] Goddard's research into the Kallikak family history began with an eight-year-old girl committed to the Vineland Training School for Feeble-Minded Children. As Goddard describes it, this girl was an illegitimate child born in an almshouse to a feebleminded mother. As a resident of the Training School, Deborah Kallikak was subjected to routine genealogical studies that came to arouse the curiosity of the research facility. Goddard and his cadre of field workers discovered not only that Deborah descended from a family long beset by poverty, but also that her family had the same surname and significant hereditary links to a wealthy and well-known family living in the same state. These two 'discoveries' offered Goddard the basis for his exploration of both branches of the Kallikak family. In terms of Deborah's impoverished lineage, Goddard sought to build upon and revise Dugdale's method of study so as to prove incontestably that environment was *not* a primary determinant of human character and experience.

Goddard believed that the quality of inherited germ plasm determined a person's mental and moral capacity, and he sought to use the Kallikak family to prove that the inheritance of "bad blood" defined a person's social potential, which "no amount of education or good environment [could] change" (Goddard 1973, 51, 53). Goddard describes his discovery of the immutability of blood-type in the Kallikak family in dramatic terms: "The surprise and horror of it all was that no matter where we traced them, whether in the prosperous rural district, in the city slums to which some had drifted, or in the more remote mountainous regions, or whether it was a question of the second or the sixth generation, an appalling amount of defectiveness was everywhere found" (16). In contrast to this "horror," Goddard offers another, distant branch of Kallikaks, a "good family of the same name," whose descendants "represent nothing but good citizenship. There are doctors, lawyers, judges, educators, traders, landholders, in short, respectable citizens, men and women prominent in every phase of social life" (16, 30). In the final pages of the study, Goddard summarizes the difference between the family's two descent lines as a stark contrast between "good citizenship" and "evil" social delinquency (102): "We find on the good side of the family prominent people in all walks of life and nearly all of the 496 descendants owners of land or proprietors. On the bad side we find paupers, criminals, prostitutes, drunkards, and examples of all forms of social pest with which modern society is burdened" (116).

Goddard explains that the central investigative question in his research was to discover the origin of the family's split—to ascertain when and from what point the family developed two separate genealogies. For Goddard's purposes, the family's split into two related but distinct lines of descent was crucial, for it allowed him to present the "good side" of the family as the genealogical norm from which the "bad side" had degenerated. To locate an originary moment for this deviation then provided Goddard both with evidence of a biological transgression (i.e., bad breeding) and with a myth of fallen grace. In proper Biblical fashion, Goddard's genealogical investigations and speculations ultimately led him to attribute fault for the seduction and the fall to a tainted woman and an unsuspecting man in the family's distant past.

The mystery of the relationship between Deborah's family and the wealthy Kallikaks was solved with the genealogical discovery of an illegitimate son born to a Martin Kallikak and "a feeble-minded girl" at the beginning of the American Revolution. According to Goddard, this Martin Kallikak joined a military company at the beginning of the war and at "one of the taverns frequented by the militia he met a feeble-minded girl by whom he became the father of a feeble-minded son" (18). In a chapter entitled "What It Means," Goddard tells the following story:

> We have here a family of good English blood of the middle class, settling upon the original land purchased from the proprietors of the state in Colonial times, and throughout four generations maintaining a reputation for honor and respectability of which they are justly proud. Then a scion of this family, in an unguarded moment, steps aside from the paths of rectitude and with the help of a feeble-minded girl, starts a line of mental defectives that is truly appalling. After this mistake, he returns to the traditions of his family, marries a woman of his own quality, and through her carries on a line of respectability equal to that of his ancestors. (50)

Although Martin Kallikak, "the scion" in the family, is faulted for his moral indiscretion (which Goddard interprets as a biological transgression), the real source of taint in this story is the "feeble-minded" girl. While Martin Kallikak becomes redeemed through marriage and social respectability, the "feeble-minded girl" attains no such redemption in Goddard's narrative. Her outcast position becomes even more acute and symbolically potent as her status moves from pariah to parasite with the succeeding generations of children born of her linage. In Goddard's study, the "feeble-minded girl" is not only the site of moral taint but also the source of proliferating biological contagion. When Goddard compares the two lines of the Kallikak family, he describes the legitimate progeny of Martin Kallikak and his wife ("a woman of his own quality"—read class) as "our norm, our standard, our demonstration of what the Kallikak blood is when kept pure, or mingled with blood as good as its own" (68). In contrast to this "norm," Goddard describes the contaminating effects of a woman of "bad stock" and her "race of defective degenerates" (12, 103). He writes, "Over and against this [pure Kallikak lineage] we have the bad side, the blood of the same ancestor contaminated by that of the nameless feeble-minded girl. From this comparison the conclusion is inevitable that all this degeneracy has come as the result of the defective mentality and bad blood having been brought into the normal family of good blood, first from the nameless feeble-minded girl and later by additional contaminations from other sources" (68–69).

Goddard's "feeble-minded girl" thus represents a prototype of the twentieth-century stereotype of the pathological unwed mother: a monstrous female whose fertility threatens middle-class familial security and whose children corrupt national identity. Goddard envisions this reproductive threat, embodied in the figure of the "feeble-minded girl," as a malignant source of contamination for two reasons. First, this "nameless feeble-minded girl" signifies a site of intra-racial instability, for not only does her socioeconomic class status disrupt dominant claims to white racial superiority, but, as her relationship with Martin Kallikak illustrates, few legal barriers separate white from white.[15] Women such as "the feeble-minded girl" not only can mix with men of higher class status, but, according to Goddard, they tend to breed. Herein lies the second reason for Goddard's fears: these women of "bad blood" have reproductive agency. Whether they are prompted by instinctive sexual drives or degenerate desires, these women exemplify "the terrible danger of procreation" when they seduce and breed according to their own impulses.[16] Within Goddard's schematized system of differentiated blood-types and tainted sexual contacts, the "nameless feeble-minded girl" is perceived as dangerous not so much for her moral transgression as for her social mobility and biological potential.

In order to understand eugenic anxieties about illicit reproduction, it is important to examine more closely Goddard's construction of the Kallikak lineage. In the lengthy quotation above, Goddard describes how the Kallikak family "of good English blood of the middle class" maintains "through four generations" a "reputation for honor and respectability." This genealogical telos becomes splintered when Martin Kallikak "steps aside from the paths of rectitude," has sexual intercourse with a lower-class woman (a woman of "bad stock"), and engenders an illegitimate line of "mental defectives." This moment of interclass and intraracial contact represents, I argue, an originary scene of perverse generation as Martin Kallikak deviates from ancestral prescription and engenders a proliferating line of social "defectives." Goddard highlights the perverse implications of this illicit genealogy when he describes Deborah Kallikak's family as the "degenerate offshoot from an older family of better stock" (1973, 16). His characterization of Deborah's family as a "degenerate offshoot" signifies not only reproductive deviation from ancestral tradition, but also the proliferating growth of an aberrant, irreverent branch of the family tree. Goddard embellishes upon the arboreal metaphor when he describes Martin Kallikak as a "scion" who warps the family tree by engendering this "degenerate offshoot." The *Oxford English Dictionary* defines the term "scion" as a descendant or an heir, and as "a shoot or twig." According to Goddard's description, Martin Kallikak is a renegade scion whose actions cause the Kallikak family tree to develop a perverted outgrowth (in the sense of deviation from normative genealogical patterns) due to illegitimate mixing.[17] This "mistake," as Goddard terms it, was partially redeemed through Martin Kallikak's return to ancestral norms, his choice to marry "a young woman of good Quaker family" (99), and his fathering of a legitimate branch of landowners and proprietors. Goddard describes this partial genealogical redemption in the following manner: "Martin Sr., on leaving the Revolutionary Army, straightened up and married a respectable girl of good family, and through that union has come another line of descendants of radically different character. These now number four hundred and ninety-six in direct descent. All of them are normal people" (29).

Goddard's representation of Martin Kallikak's having "straightened up" is both telling and resonant. His imagery points not only to a realignment and normative rehierarchization, but also to a prior moment of developmental deviation that engendered radical—and proliferating—genealogical alteration. Goddard's central focus in his study is on the social presence and generative potential of perverse "offshoots," and the anxiety associated with such deviation exemplifies what Dollimore terms "the paradoxical perverse" (1991, 121). Dominant definitions of perversity in Western culture have developed around hierarchical binary oppositions of natural/unnatural, true/false, inside/outside, male/female, etc. Dollimore argues that these binaries are constructed so that the second term represents not only an inferior, subordinate version of the first, but also a perversion of the first term's truth and authenticity. As discussed earlier in this essay, Dollimore shows that the paradox of this construct exists in the fact that the second term in the binary is understood as having deviated from, and thus originally having developed out of, the telos of the dominant term (108–13). "[T]he shattering effect of perversion," Dollimore writes, "is somehow related to the fact that its 'error' originates internally to just those things it threatens" (121). The central strategy in this definitional enterprise is the displacement of conflict and disorder from the dominant onto the subordinate, the latter of which is then defined as outside the realm of the natural and the normal. But, Dollimore points out, this subjugating process all begins with disorder within the dominant, and, as he insists, this instability marks the site not only of dominant displacement but also the possibility for internal fracture: "deviation and perversion are what the dominant defines itself against, yet simultaneously deviation and perversion emerge from within, are produced by and displaced from the dominant. There occurs a 'splitting of the true'" (113).

In Goddard's story, the "splitting of the true" occurred quite overtly at the historical moment when Martin Kallikak deviated from ancestral tradition and reproduced an illegitimate child with "the nameless feeble-minded girl." In his study, Goddard constructs Martin Kallikak as a type of scientific control in a hereditarian experiment: Martin represents the stable testing means while the characteristics of his progeny are assessed in relation to the supposed "nature" of his breeding partners. Good reproduction is signified by internal union (the inbreeding of Martin and a woman of his own class), while perverse reproduction originates in the illegitimate union between Martin and the lower-class "feeble-minded girl." Goddard makes the identity of the mother a determining factor in the familial split. He writes, "The Kallikak family presents a natural experiment in heredity. A young man of good family becomes through two different women the ancestor of two lines of descendants—the one characterized by thoroughly good, respectable, normal citizenship, with almost no exceptions; the other being characterized by mental defect in every generation" (116). While Martin is redeemed (and sociosexual boundaries are again closed off) when he marries "a woman of his own quality" (50), blame for the genealogical deviance is displaced onto the feebleminded girl who is pathologized as the origin of generations of socially deviant children. According to Dollimore, this pattern of deviation, displacement, and reconsolidation is both ruthless and routine for the (re)legitimation of dominant authority:

> [A]n irruption within the dominant destabilizes the binary oppositions legitimating that order. The binary is then in part restabilized through renewed control of those signified by its inferior term, who are typically identified as inverting, perverting, or deviating from the prevailing order, and in the process have displaced onto them responsibility for the disruption occurring elsewhere. (Dollimore 1991, 112)

While blame is laid on the feebleminded girl in Goddard's narrative, the "irruption within the dominant" that preceded her being made monstrous occurred with Martin Kallikak's reproductive "mistake" and the racial transgressions signified by his deviation. Martin's having "dallied" with a purportedly "feebleminded" member of the lower class exposes the instability of class boundaries and bloodline insularity, but this exposure becomes veiled by the exclusion and denigration of a new line of "mixed" persons.

The attention to heredity and the belief in the immutability of germ plasm placed the eugenicists in the difficult position of having to contend with unwanted familial connections. Distinctions between good and bad, normal and deviant remained blurred by the proximity and affinity of blood relations. While illegitimate progeny were denied familial legitimacy and, as in the case of the Kallikaks, excluded from access to social power, their blood connection to the "good" family existed and persisted nevertheless. In specifically demonizing "feeble-minded" women and labeling them as sexual contaminants, eugenicists thus developed a powerful and effective means to identify an alien source of "bad blood" and to reconstitute the dominant inside/outside boundary. By displacing blame for sexual and social transgression onto the reproductive bodies of impoverished women, eugenicists sought to reify dominant class and race hierarchies, yet their diagnostic agenda reveals the selectivity of scientific claims of white racial superiority. While eugenicists strove to warn the American public against undetected threats to white racial progress, the popularity of their theories also made visible the ways in which white racialism can fracture at the point of class integration.

Notes

1. The epigraph to this essay is excerpted from an article quoting Amos Butler in the social welfare magazine, *Charities* (Butler 1903, 599). The second quotation from Butler appears in his 1901 article, "A Notable Factor of Social Degeneration" (Butler 1963, 339). My knowledge of the second quotation is indebted to Mark Haller's research on American eugenic philosophies (Haller 1963, 27).

2. White single mothers who came under study by eugenicists were believed to have flaws in their genetic makeup, defined as weak "germ plasm," which not only influenced (or determined) their sexual tendencies, but were also transmittable to their children. For an example of this pseudo-scientific theory about defective "germ plasm" and its hereditary implications, see Davenport 1909.

3. It is difficult to assess the number of women diagnosed as feebleminded and subject to punitive social measures. Institutionalization, sterilization, denial of economic aid, and the removal of children from the home were only some of the measures perpetrated against impoverished women under the guise of mental health care and social reform. It is also difficult to determine accurately the mental and emotional condition of those women diagnosed as "mentally defective," for not only do state records reflect the perceptions and biases of those in charge, but (some) medical standards concerning mental health and mental illness have changed over the years. State records on sterilization procedures offer one way to assess the size and breadth of the eugenics influence on the mental health movement. In *The Surgical Solution,* Philip Reilly describes the extent and the implications of America's history of involuntary sterilization: "During the first six decades of the twentieth century, more than sixty thousand mentally retarded or mentally ill individuals, most of them residents of large state institutions, were sterilized for eugenic reasons. Even the most accommodating view of sterilization programs must recognize that at best only a tiny fraction of such persons consented to the surgery. Indeed, hundreds and possibly thousands of people, most of them women, were never told that they had been sterilized. It is extraordinary how many 'appendectomies' were performed at some state homes for the retarded in the 1920s and 1930s" (Reilly 1991, xiii).

4. Any conceptualization of moral regulation in early twentieth-century America cannot be separated from the racial debates that dominated the cultural landscape at the time. Two different, but certainly interrelated, sites of cultural contestation developed at the turn of the century that greatly influenced twentieth century racial politics. First, the failure of Reconstruction efforts to establish and secure civil rights for newly freed African Americans and the concomitant backlash of Jim Crow laws established a new severity in black-white segregation mandates, and even greater pressure on scientific racialism to establish fundamental biological differences between the races. Second, the geographic composition of immigrants from Europe began to change in the 1880s. Different from the national origin of earlier European-born immigrants (deriving predominantly from Germany, Great Britain, Scandinavia, France, Switzerland, and the Netherlands), the "new immigrants," as they came to be called, originated from eastern and southeastern Europe. By tracing early American democratic ideology to idealized notions of liberty and democracy allegedly developed by northern European Anglo-Saxon and Teutonic tribes, American nativists established Anglo-Saxon ancestry as a criterion for true American identity. According to this logic, " 'old Americans'—those of English and northern antecedents—were the 'real Americans.' " This nativist quotation about "real Americans" is excerpted from Gossett 1963, 122. See also Higham 1992.

5. The idea of "germ plasm," an early genetics theory developing out of August Weismann's work on biological determinants, became popular in twentieth-century eugenic arguments for reproductive regulation. "Germ plasm" was the inherited organic substance that carried character traits, including mental capacity and criminal potential, that were believed to be unaffected by environmental influence (Kevles 1995, 18–19, 70–71). With classifications of feeblemindedness, more attention was paid to the mental incapacities of *individual* white persons because dominant society was already obsessed with proving the mental, physical, and social inadequacies of *all* black persons. Both groups were

deemed biologically flawed, but white "mental defectives" were seen as instances of reproduction gone-wrong, whereas black persons, as a race, were defined as inherently inferior. According to turn-of-the-century white supremacist discourse—often directed toward black political disenfranchisement—the genetic composition of "black blood" determined a limited mental capacity for African Americans. See for example the intelligence assessments offered by the prominent white sociologist Howard Odum in 1910 in *Social and Mental Traits of the Negro* (Odum 1968). Because racist discourse defined blacks as always already mentally inferior, it is not surprising that scarce legislative attention was paid to African American mental health in the first few decades of this century. In a study of the treatment of "black feeble-mindedness" in the South, Steven Noll shows that legislative concern about "mentally defective Negroes" existed, but little was done to address the issue. Nearly all state institutions were designated for whites only, and very few funds were allocated to address African American mental health. Ironically, this racist neglect somewhat shielded African Americans from the *legal* sterilization procedures performed on "feeble-minded" persons in state institutions (illegal procedures, such as lynching, provided another, related, type of social control). Noll writes, "before 1950, blacks comprised only 23 percent of North Carolina's sterilizations at a time when they made up over 28 percent of the state's population" (1991, 143). However, it is important to recognize that this ratio changed significantly in the 1960s when dominant discourse began to focus on a stereotype of black single mothers as the cause of rampant social ills. In this bigoted and brutal political climate, persons victimized by involuntary state sterilization changed in aspect. Noll writes, "North Carolina continued to sterilize individuals for eugenic reasons [in the 1960s]. Of the 863 persons eugenically sterilized by the state of North Carolina from 1962 to 1966, 64 percent were black" (151 n. 61).

6. *Oxford English Dictionary*, 1979, "degeneration"; my emphasis.

7. Dugdale 1975, 60. I use the word "unregulated" here in the sense of both cultural and juridical nonconformity. Not only did the reproductive history of the Juke family, *as described by Dugdale*, not conform to normative middle-class moral mandates, it also represented a direct affront to state prohibitions on behavior such as prostitution, illegitimacy, incest, and pauperism.

8. I derive the notion of the "governed hearth" from Michael Grossberg's work on nineteenth-century family law and dominant conceptions of a bounded, discrete domestic unit. In *Governing the Hearth*, Grossberg chronicles the change from colonial America's understanding of the family as a reciprocal part of the larger community to a more partitioned, privatized legal definition of the family during the nineteenth century. This change entailed a shift from patriarchal domestic governance at the level of the family to state governance of members of the family via the means of public legislation and juridical decisions. Grossberg argues that nineteenth-century domestic law became a central means for the state to "establish the boundaries within which families formed and lived" (1985, 29). I would add that domestic legislation also became a means to enact normative mandates for family structure and gender roles, and to justify state intervention in instances of non-conforming, thus criminal, domestic and sexual relations.

9. Dugdale's description might serve as ironic commentary on nineteenth-century racist idealizations of whiteness if the Juke family had not been made a prototype for the eugenic campaign against "the feeble-minded" which, in reality, became a vicious form of class oppression.

10. Charles Davenport states this in his preface to *The Jukes in 1915* (Estabrook 1916, iii). Davenport attributes this "bad germ-plasm" to criminals who emigrated from Europe during the colonial period. This theory of criminality and bad blood was specific to Davenport and the eugenic agenda that motivated the 1915 follow-up report on the Juke family. Dugdale knew very little and had little to say about the first Juke he traced, Max.

11. See Dugdale 1975, 19–28, 111.

12. My understanding of changes in the American eugenics movement is indebted to Edward Larson's succinct explanation of how scientific developments at the turn of the century affected hereditarian theory. Larson explains how the work of August Weismann, a German cytologist, challenged neo-Larmarckian theories of the inheritability of acquired characteristics, and how the rediscovery of Mendelian biology suggested that unit characteristics were immutable and inherited according to regular, calculable patterns (intelligence was believed to be one of these unit characteristics). Unit characteristics thus became conceived as all-determinant and unaffected by such "misguided" reform efforts as charity relief, education, and criminal rehabilitation (Larson 1995, 19–20).

13. "Kallikak" was a pseudonym given by Goddard that derived from the Greek words, "kallos," meaning beauty, and "kakos," meaning bad (Smith 1985, 12). In his study of the Kallikak report, *Minds Made Feeble: The Myth and Legacy of the Kallikaks*, J. David Smith seeks to examine the social and academic effects of Goddard's study and to develop a more accurate genealogical history for the Kallikak family. While I have benefited from his research, my work differs from that of Smith in that I explore not just how Goddard chose to portray the Kallikak family history, but also why it is that the reproductive potential of impoverished white women, specifically, played such a central role in eugenicists' fears about racial vulnerability.

14. Goddard's study became highly popular reading and reached a wide audience. *The Kallikak Family* was a bestseller, went through several printings, and was taught in many academic settings (Smith 1985, 5). Goddard was also approached by a Broadway drama agent concerning rights to the book. In *Minds Made Feeble*, Smith includes a letter written to Goddard by Alice Kauser, a Broadway agent, and then Goddard's response to her. Goddard wrote to Kauser, "I am sure that we should have to be assured that the play [you suggest] would be one that would carry the moral lessons which the book is intended to convey. We would not consent to its being dramatized for any other purpose. Now, whether this can be done and still make it attractive and a success, you will know much better than I" (Smith 1985, 63).

15. Here I am discussing the scarcity of *explicit* rules on intraracial mixing as compared to other, far more stringent legal and cultural prohibitions on interracial mixing. Goddard's warning to men of the upper classes to beware the dangers of class-mixing illustrates the far less punitive social responses to such sexual transgressions. Goddard cautions: "The career of Martin Kallikak Sr. is a powerful sermon against sowing wild oats. Martin Kallikak did what unfortunately many a young man like him has done before and since, and which, still more unfortunately, society has too often winked at, as being merely a side step in accordance with a natural instinct, bearing no serious results. . . . Now that the facts are known, let the lesson be learned; let the sermons be preached; let it be impressed upon our young men of good family that they dare not step aside for even a moment" (1973, 102–3).

16. Goddard 1973, 106; see also 65–67, 27.

17. The *OED* also defines a scion as "a slip for grafting." It is interesting to relate Goddard's fascination with Mendelian plant-grafting experiments to his anxieties about class and racial miscegenation. In the case of human reproductive "mixing," as opposed to plant grafting, eugenic anxiety centered around the fact that hybridity was not checked by sterility.

References

Butler, Amos. 1903. "Feeble-minded Women in Indiana." *Charities* 11, no. 25: 599–600.

———. 1963. "A Notable Factor of Social Degeneration." 1901. *Proceedings of the American Association for the Advancement of Science* 50: 339. Quoted in Mark Haller, *Eugenics: Hereditarian Attitudes in American Thought*. New Brunswick, N.J.: Rutgers University Press, 27.

Davenport, Charles. 1909. "Influence of Heredity on Human Society." *Annals of the American Academy of Political and Social Science* 34 (July): 16–21.

Dollimore, Jonathan. 1991. *Sexual Dissidence: Augustine to Wilde, Freud to Foucault*. Oxford: Clarendon Press.

Dugdale, Richard. [1877] 1975. *"The Jukes": A Study in Crime, Pauperism, Disease and Heredity*. Reprint, 1895 edition in Foundations of Criminal Justice Series. New York: AMS Press.

Estabrook, Arthur. 1916. *The Jukes in 1915*. Preface by Charles Davenport. Washington, D.C.: Carnegie Institution of Washington.

Goddard, Henry. 1912. "Social Investigation and Prevention." *The Survey* 27, no. 22: 1852–56.

———. [1912] 1973. *The Kallikak Family*. Reprint, in Classics in Psychology. New York: Arno Press.

Gordon, Linda. 1994. *Pitied But Not Entitled: Single Mothers and the History of Welfare, 1890–1935*. New York: Free Press.

Gossett, Thomas. 1963. *Race: The History of an Idea in America*. Dallas, Tex.: Southern Methodist University Press.

Gould, Stephen Jay. 1977. *Ever Since Darwin*. New York: W. W. Norton.

Grossberg, Michael. 1985. *Governing the Hearth: Law and Family in Nineteenth Century America*. Chapel Hill: University of North Carolina Press.

Haller, Mark. 1963. *Eugenics: Hereditarian Attitudes in American Thought*. New Brunswick, N.J.: Rutgers University Press.

Higham, John. 1992. *Strangers in the Land: Patterns of American Nativism, 1860–1925*. 2nd ed. New Brunswick, N.J.: Rutgers University Press.

Johnson, Alexander. 1909. "Race Improvement By Control of Defectives (Negative Eugenics)." *Annals of the American Academy of Political and Social Science* 34 (July): 22–29.

Kevles, Daniel. 1995. *In the Name of Eugenics*. Cambridge, Mass.: Harvard University Press.

Larson, Edward J. 1995. *Sex, Race, and Science: Eugenics in the Deep South*. Baltimore, Md.: Johns Hopkins University Press.

Mink, Gwendolyn. 1990. "The Lady and the Tramp: Gender, Race, and the Origins of the American Welfare State." In *Women, the State, and Welfare*, edited by Linda Gordon. Madison: University of Wisconsin Press.

Noll, Steven. 1991. "Southern Strategies for Handling the Black Feeble-Minded: From Social Control to Profound Indifference." *Journal of Policy History* 3, no. 2: 130–51.

Odum, Howard. [1910] 1968. *Social and Mental Traits of the Negro*. Reprint, in Studies in History, Economics, and Public Law 99. New York: AMS Press.

Reilly, Philip R. 1991. *The Surgical Solution: A History of Involuntary Sterilization in the United States*. Baltimore, Md.: Johns Hopkins University Press.

Smith, J. David. 1985. *Minds Made Feeble: The Myth and Legacy of the Kallikaks*. Rockville, Md.: Aspen Systems Corporation.

"Tourists Accommodated" with Reservations: Dorothy Canfield's Writings, Vermont Tourism, and the Eugenics Movement in Vermont

Julia C. Ehrhardt

RECENT SCHOLARSHIP IN WOMEN'S LITERATURE AND CULTURAL studies has resurrected academic interest in Dorothy Canfield Fisher's life and work. Studies have analyzed Canfield Fisher's tenure as an influential judge for the Book-of-the-Month Club and documented her many humanitarian efforts in the areas of war relief, adult education, and race relations.[1] Yet, despite the many accolades she received for these humanitarian endeavors, she wished to be remembered for the novels, short stories, and articles that made her one of the most popular writers in the first half of the twentieth century. As she passionately insisted in a letter to David Baumgardt on 13 February 1957, "My efforts to be a good citizen . . . really should not be given so much attention that my books, to which I have given the very core of my heart and mind, should be pushed into the background" (Fisher Papers). Despite Canfield Fisher's concerns about her legacy, scholar Mark Madigan maintains that she should not have worried about the possible demise of her literary reputation. Praising the wide range of topics Canfield[2] explored in her fiction (including women's rights, love, war, greed, and race relations), Madigan correctly observes that she was "gifted with an ability to distinguish the enduring from the merely topical," concluding that her work is "often startling in its contemporaneity" (1996a, 11).

That a new generation of scholars is beginning to acknowledge the relevance of Canfield's fiction to present-day social concerns proves Madigan's assertions. In attempts to demonstrate the continued relevance of Canfield's writings, though, critics have given short shrift to one of their most compelling historical contexts: the development and growth of the Vermont tourist industry during the first four decades of this century. At the same time Canfield was establishing herself as a best-selling author on the American literary landscape, Vermont reinvented the place it occupied on the national map, transforming itself from an economically depressed farming region into a nationally popular vacation destination. Initially, this correlation seems merely coincidental, but the connection becomes clearer when we recognize how many of Canfield's works are set in Vermont and feature tourism as a central theme. This literary-historical relationship is significant for a more troubling reason, however, because of the writer's participation in the dubious activities of the Vermont Commission on Country Life, an organization intimately related to the eugenics movement in Vermont. When Canfield's writings about Vermont and its tourism are interpreted using a critical perspective that incorporates the history of the state's eugenics movement, this author emerges not as a writer ahead of her time, but as one inextricably enmeshed in one of the most important—and discomfiting—periods in United States history. Moreover, investigating the historical connections between the apparently unrelated topics of eugenics and tourism as Canfield articulated them attests to the pervasiveness of eugenics ideology in America in the first half of the twentieth century.

It is not surprising that Dorothy Canfield (1879–1958) embraced tourism as a central theme for her writing since she traveled extensively in the United States and Europe from the time she was a child. Though she had seen European capitals and had watched seals in San Francisco by the time she was a teenager,[3] Canfield's favorite journeys were the many trips she took to her paternal grandparents' farm in Arlington, Vermont. In addition to horseback riding and farm chores, she loved to listen to her relatives tell stories about their pioneer ancestors. Canfield memorialized these summer escapades in her earliest literary efforts: letters home to her parents and several school compositions.[4] As she entered young adulthood, Canfield continued to write to her parents about her travels, and her first professional publication resulted when she sent a letter she had written to her father about a Holy Week trip in Spain to the editor of *The New York Times*. After the piece appeared in print, Canfield decided to pursue a career as an author and, while touring Europe in the summer of 1905, began to write a novel about a New England family vacationing in Scandinavia.

In 1907, the same year Henry Holt published Canfield's "Norwegian story," (Washington 1982, 58), Canfield married James Redwood Fisher. The couple then made the risky decision to relocate from New York City to Arlington, the bride's beloved ancestral village. Despite the writer's distance from the nation's literary epicenter, her career skyrocketed. By 1910, Canfield had completed the manuscript for her second novel and was trying to satisfy the overwhelming demand for her short fiction. In addition, she had agreed to translate into English the writings of the Italian educational philosopher Maria Montessori. Before leaving for a trip to Europe, Canfield approached her editor, Alfred Harcourt, with an idea for yet another book: a collection of previously published stories about "Hillsboro," a fictional Vermont village modeled upon Arlington.[5]

Canfield considered her "Hillsboro stories" as among her best literary works, and given the historical context from which they arose, her desire to republish them in book form made perfect sense. At the same time the Hillsboro stories were captivating readers of popular periodicals, Vermont officials were desperately struggling to revise the state's dubious reputation as the preeminent symbol of New England's decline. As several historians have established, in the decades following the Civil War, Vermont's economy began to founder as railroad networks enabled Midwestern dairy farmers to compete with their Vermont counterparts (Rebek 1997, 15) and most of the state's industries relocated to southern New England's urban centers (Brown 1995, 137). A mass exodus from the Green Mountains ensued as Vermonters left their homes to seek more profitable work elsewhere (Rebek 1997, 16). Cultural critics in periodicals such as *Scribner's*, *Atlantic Monthly*, and *The Century* bemoaned Vermont's "decay" as symptomatic of northern New England's financial struggles (22) and lamented the region's growing reputation as a cultural cesspool (Brown 1995, 140).

As historian Dona Brown explains, enterprising tourist boosters hoping to jumpstart local economies began to market the "small towns, rural virtues, and ethnic purity" that could still be found in Maine, Vermont, and New Hampshire as antidotes to the hustle and bustle of city life (1995, 140). Foremost among the three states, Vermont capitalized on its potential as a prime vacation destination. According to Brown, Vermont tourism promoters followed the example of New England local color writers by using print to advertise the nostalgic aura of the region (153). Brochures and pamphlets promised an easy panacea to readers weary and worn down by the "mobility, greed, and individualistic competition" (148) that had come to characterize modern life. Articles and illustrations in these publications promised that taking a vacation in the Vermont countryside would guarantee a return to peace, contentment, community cohesiveness, and republican simplicity (148) for anyone craving a "healthy respite" from the stresses of city living (Calder 1997).

Although scholars have not considered Canfield's *Hillsboro People* as one of the author's literary promotions of Vermont,[6] the content of the volume suggests that the writer was well aware of the state's publicity program and intended her book to serve as another advertisement for Vermont tourism. In the preface, Canfield indeed affirms that she composed the stories to enable readers who could not make an actual vacation trip to her town to take one in "canned" form (1915, 9). Her tales about "Virginia," a world-weary traveler who finds new meaning in her life during her "Old Home Week" trip to Hillsboro, a farm wife named Matey Prentiss who attempts suicide when her son urges her to move to New York City, and a resourceful spinster named Abigail Warner who revives the economy of her dying mill town depict Vermont as an ideal vacation spot. In addition to individualism, ingenuity, and integrity, Canfield invests her Hillsboro characters with devout loyalty to tradition, devotion to place, and respect for the past. Thus, they literally represent another ideological facet of the state that tourism boosters highlighted in their promotional materials: that Vermont, along with the rest of New England, was the geographic source of the nation's "political courage, personal independence, and old-fashioned virtue" (Brown 1995, 153).

In addition to admirable characters and breathtaking scenery, the lack of ethnic diversity in Hillsboro, as well as Canfield's disparaging descriptions of its few foreign citizens, would have definitely appealed to readers seeking relief from the increasing tides of immigration that were changing the complexion of many eastern cities. Subtle details in the Hillsboro stories indicate that Canfield imagined Vermont as a respite not only from threatening industrial trends, but also from threatening people. A closer reading of "Finis" suggests that it is not the small size of her son's New York apartment that leads Matey Prentiss to attempt suicide, but the ethnic intermingling to which she will be exposed if she moves there. As she recounts, her son Hiram has married a "queer Dutch wife" and has subsequently been transformed into a "half Teutonic" alien. Her fears about his devolution are borne out in her grandsons, whose "German-American" appearance, "loud" voices, and effusive "affection" shock her native sensibilities (1915, 235). Canfield also presents dubious portrayals of immigrants in the story "A Drop in the Bucket" (in which the primary antagonist is a "dirty" socialist "Canuck shoemaker" with "hot blood")[7] and in the tale "Adeste Fideles." In this piece, when spinster Abigail Warner reveals her ingenious plan to reopen a mill in her dying town, neighbors raise the possibility that the "greasy millhands" will eventually destroy their "Paradise" (345).

Canfield's uncomplimentary descriptions of foreigners and immigrants as well as other pointed details in the stories (for example, Abigail Warner's hobby of breeding plants) evoke the eugenic explanation of "New England's decline" that had blossomed in the 1920s and reached full flower by the 1930s. Ever-increasing numbers of immigrants settling in New England incited a social panic in the region, as academicians, politicians, and concerned citizens convinced themselves that a "race war" was imminent. As Daniel Chauncey Brewer explained in his 1926 manifesto, *The Conquest of New England by the Immigrant*, a Darwinian struggle with the "tough and resisting qualities" of the region's landscape had resulted in the evolution of a "master race" of Yankees (30). However, due to declining birthrates among native-born women and the overwhelming exodus of native New Englanders to other regions of the country, the once formidable native stock had attenuated (345). In an evocative passage, Brewer symbolized the Yankee race as a tree whose "inner trunk" was "crumbling" (299) and subsequently, in a masterfully mixed metaphor, claimed that foreigners' tendency to reproduce "without restraint" (345) constituted a "sponge" that would soon "wipe" the Yankee "off the face of contemporary history" (173). Though Brewer worried about the fate of the entire region, he specifically identified Vermont as a traditional "Yankee stronghold" that was "losing rather than winning ground" in the battle for racial superiority (346).

Brewer's book and similar texts served as a desperate draft notice specifically directed to Vermont natives who could serve as soldiers in the race war. The leader of the troops, according to historian Kevin Dann, was Henry Perkins, a professor of zoology at the University of Vermont. Perkins was an ardent advocate of eugenics, the "science" of improving human "races" by controlling hereditary factors through selective mating (Dann 1991, 1). Like his fellow eugenicists, Perkins believed that certain racial "stocks" possessed the most desirable human characteristics and therefore should be encouraged to procreate, while members of less genetically gifted groups should be prevented from mating (a concept known in academic parlance as "negative eugenics"). In 1925, under the auspices of the president of the University of Vermont, Perkins organized a comprehensive "Eugenics Survey" of the state, which he designed with hopes of preserving the virility of Vermont's native "breeding stock." After analyzing the pedigrees of six thousand people, Perkins concluded that the "hereditary trends" in the "old families" of Vermont were most valuable to the state (1925, n.p.), while the "depravity, immorality, and loose living" in "gypsy" and "pirate" populations made them "a constant menace" to Vermont's welfare (Dann 1991, 6). He leveled similar charges against French Canadians, the largest immigrant group in Vermont (15).[8]

Perkins's results reinforced the popular belief that Vermont's genetic arsenal had been drastically depleted, and he fully expected that as a result of his studies, the state would adopt a policy of mandatory institutionalization or sterilization for those persons he believed would contaminate the state's gene pool (1991, 9). In order to further ensure the purity of the Vermont pedigree, Perkins embarked on another research project in 1927: to take social stock of the state by investigating every aspect of Vermont culture, with the ultimate goal of protecting traditional Vermont life and customs from the intrusions of immigrant interlopers. Two hundred Vermont citizens volunteered to assist Perkins in the project, which was known as the Vermont Commission on Country Life. One of these volunteers was Dorothy Canfield.

The extent of Canfield's involvement with the Commission has proved a contentious issue among scholars, no doubt due to the organization's direct ties to the Eugenics Survey. Canfield biographer Ida L. Washington does not discuss the writer's involvement with the group, and Madigan asserts that although Canfield did participate in the activities of the Commission, she "never advocated the biological engineering supported by the eugenics movement" (1996b, xv). As Hal Goldman acknowledges, the author dedicated much of her writing and her life to "combating intolerance, bigotry, and authoritarianism" and thus the extent of her involvement with the organization must be scrutinized with care (1997, 140). Evidence suggests, though, that despite her lifelong efforts to fight racial discrimination and other social injustices, Canfield must have regarded the Commission's project as a crucial endeavor. In 1928, despite enormous demands on her personal time, she accepted a position on the Committee for the Preservation of Vermont Literature and Ideals, and in 1932 was elected to the Commission's executive board.[9]

In 1931, the Commission presented its early findings in a book entitled *Rural Vermont: A Program for the Future*. It was the duty of Canfield's committee to formulate a plan that would preserve in future mating populations the "distinctive" values that made Vermonters such "hardy, independent, liberty-loving, brave and individualistic people" (Two Hundred Vermonters 1931, 372). The eleven-member committee drafted a report outlining measures the state could adopt to preserve its unique architecture, music, and literature. It also proposed a quite novel suggestion for improving the state's breeding stock: that Vermont should actively recruit "authors," "artists," and "others in the same general classification" to vacation there. These sojourners could subsequently be convinced to relocate to Vermont permanently, thus transplanting themselves into Vermont's pathetic genetic seedbed (385).

Although the committee's report did not pack the same type of sinister punch as did Perkins's plans for social undesirables, it advocated that the state adopt an equally discriminatory program of social engineering. As Canfield's committee reckoned, some types of vacationers (college professors, lawyers, and writers, for example) were "far more valuable to Vermont" than others (Two Hundred Vermonters 1931, 380)—most specifically, the hordes of automobile tourists who flocked to Vermont every summer. Canfield's committee viewed such transient travelers in the same way the zealots of the Vermont "race war" saw immigrants: as uninvited invaders who threatened to destroy the fundamental fabric of Vermont life. At the turn of the twentieth century, Vermont vacationers had been a small, self-selected class of people; middle- and upper-class urban professionals who possessed enough time to take a leisurely train ride and enough money to spend for hotel rooms or summer homes. By the 1930s, as historian Hal Goldman explains, the demographics of Vermont tourism had changed. Due to the increasing affordability of automobiles, Vermont vacations were becoming more financially feasible. Instead of booking expensive lodging, Depression-era automobile travelers could rent rooms from rural farm families eager to meet people from other parts of the country (Two Hundred Vermonters, 124).[10] Therefore, "almost anyone of any social or ethnic background" who owned a car could invade the state that had long prided itself as a "refuge" from the tumultuous upheavals in modern life (Goldman 1997, 149).

As her writings from the period show, Canfield feared that these inappropriate tourists would wreak irreparable havoc in Vermont. In 1932, she authored *Tourists Accommodated*, a play that she wrote specifically for newly forming community theatre groups to perform. Because it depicts the misadventures of a poor Vermont family who takes in automobile tourists, the play has been read as a farce (Goldman 1997, 141), but when read in the context of the Commission's activities, it is obvious that Canfield intended to dramatize serious scenarios in her ironically-titled comedy. The play opens when Sophia Lyman, a Vermont farmer's wife, sadly informs her daughter Lucy that the family cannot afford to pay her tuition to the state teacher's college. When nosy next-door neighbor Aunt Nancy Ann learns of Lucy's predicament, she suggests that the Lymans consider renting out rooms to automobile tourists to earn extra money. As soon as Aunt Nancy Ann proposes the idea, Sophia immediately dismisses it: "Why, seems as though it would give me heart failure to have a lot of strangers come in and me try to make 'em pay for sleeping in our beds. How would I know how much to ask 'em for anything? Suppose they didn't pay me the morning after? Suppose they up and said some of us stole something from 'em? Suppose some French Canucks come along that don't speak English?" (21). As she contemplates the potential problems the tourists' children might cause, she dismisses the idea, but her commitment to Lucy's education ultimately causes her to consent to the plan.

Sophia's anxieties about city people attempting to get something for nothing, foreigners who do not speak her native tongue, thieves masquerading as tourists, and most threatening of all, the offspring of these groups reflect the Commission's anxieties about mobile tourism—all of which are borne out in the subsequent action. When visitors begin to darken the Lyman doorstep, a version of each scenario that Sophia has feared comes to pass: a Scotch-American family with obnoxious children refuses to pay Sophia the full price for lodging, another woman forcibly purchases one of the family's treasured heirlooms, and a French-speaking Canadian frightens the Lymans out of their wits. It is crucial to observe here that, while Canfield successfully overturns the demeaning stereotypes about country bumpkins upon which regional comedies traditionally rely, she replaces them with other reductive images about ethnic groups. Scotch thrift is rewritten here as deliberate swindling, and the loud whiny woman who wrestles another tourist for purchasing rights to the Lymans' antique cabinet evokes the anti-Semitic stereotype of the bargaining Jew. When the French-speaking Canadian couple arrives at the farm (after Lucy proudly asserts that she can tell what any group of tourists will say before they even knock on the door), Aunt Nancy Ann assumes that they are "crazy folks escaped from the asylum." Sophia corrects her by admonishing, "They're not crazy! They're French!" (58). But even though she can recognize the couple's mental stability, she refuses to acknowledge that the visitors are tourists in need of accommodation. The meaning of this scene is clearly evident despite the language barrier: in Vermont, accommodation is a privilege, not a right.

Over the course of the play, Sophia's reservations become reality as the steady stream of mobile tourists takes its toll on her family's values. Most significantly, in the penultimate scene of the play, Canfield explodes the notion that Vermont children would benefit from meeting cosmopolitan tourists when the Lymans' youngest son Phillip announces that he is going to run away from home to join a gypsy tribe (69). The heart failure Sophia anticipated in the opening scene appears imminent when, just after Phillip makes his devastating announcement, she answers a knock on the door and encounters a penniless transient carrying a bulky knapsack. The hobo initially embodies Phillip's potential degeneration, but when the tramp states that he is an artist from Vermont who has been seeking inspiration in the mountains, Sophia welcomes him inside and Lucy runs to get him some food. Notably, these hospitable gestures stand in sharp contrast to those the Lymans bestow upon stranger strangers.

The subsequent action proves that Canfield deliberately intends the final tourist to be a "desirable" visitor in her reckoning, though he resembles a gypsy tramp. After the rest of the family goes to bed, the artist has a long talk with Phillip and convinces him that although he has the right to leave his family, the boy also possesses the moral responsibility to remain at home. As the artist explains, Philip's presence at the family farm represents the most significant payoff of any tourist's visit, for when visitors see him, they get what they paid for: an increased appreciation of the value of "staying put" in the place where they "really belong" (78, 79).

The Lymans discover the true import of the artist's speech at the end of the summer when they count their money. After settling accounts, the family happily learns that their venture has more than paid off: in addition to Lucy's tuition they have earned enough money to pay the property taxes on their farm. It is significant, therefore, that when Lucy remarks that she cannot believe the family's good fortune, she makes no reference to money. Instead she exclaims, "Doesn't it seem too good to be true, to have everybody gone and to be by ourselves?" (90). With these lines, Canfield insinuates that the Lymans have earned an unexpected dividend in the form of the priceless lesson they have learned from the experience: that there is no place like a home—or a state—free from "undesirable" visitors and thus in an ideal world, people should "stay put" in their proper places.

Tourists Accommodated was not the only expression of Canfield's reservations about inappropriate visitors to Vermont. In 1932, the same year she wrote the play, she composed a pamphlet titled "Vermont Summer Homes" that was published by the state's tourist publicity bureau. The pamphlet advocated the same message as the Vermont Commission on Country Life: that if the state could not raise sufficient numbers of natives fresh, it would take them canned. But the brochure was implicitly written along the same lines as Canfield's play, as it made clear that some visitors were more desirable than others. In what she termed a "special invitation" to people seeking to purchase summer or year-round dwellings in her home state (1949, 3), Canfield maintained that a Vermont home would benefit newcomers immeasurably by enabling them and their offspring to establish "tap roots" in stable communities. However, as she asserted in the brochure, her invitation was intended only to those persons who shared a "common ground" with the state's natives; people who already possessed "character, cultivation, and good breeding" (4) and who could thus successfully transplant themselves into Vermont soil. Notably, Canfield's invitation specifically excluded people who "manufacture, or buy and sell material objects or handle money"(4).

Canfield's writings thus represented a most discriminating brand of residential recruitment. Yet, during the 1930s, publicity articles and travel guides celebrated the author as a state patriot along with such venerable Vermonters as Ethan Allen and Thomas Crittenden. In *Let Me Show You Vermont*, Charles Edward Crane declared that "the latchstring of both her home and her heart is always out" and that she eagerly tolerated interruptions to welcome "distinguished and undistinguished" visitors alike (1937, 273). However, the experiences of one such sojourner, the writer Anzia Yezierska, belie Crane's encomiums. Yezierska, the author of *Bread Givers* and other novels depicting the struggles of Jewish immigrants in Manhattan's Lower East Side ghetto, decided to move from New York City to Arlington, Vermont, in 1931. A self-described member of an "uprooted race"[11] and an ardent admirer of Canfield's fiction, Yezierska believed that by relocating to her idol's home state she would find the happiness depicted in the writer's "peaceful tales of small-town folk . . . a world where people were rooted in the hills and valleys of the countryside around them" (Yezierska 1987, 199–200).

Although *Tourists Accommodated* and "Vermont Summer Homes" were not published until months after Yezierska had arrived in Arlington, as a literary professional desperately seeking to escape from the material trappings and urban distractions that had adversely affected her art, she perfectly fit Canfield's profile of an ideal transplant. A letter dated "Tuesday Morning," which Yezierska wrote to Canfield, indeed attests that the newcomer initially found in Vermont the paradise the Hillsboro stories had promised: "For the first time since childhood, I. . . . slept a dreamless sleep. . . . I awoke to the song of birds and the fresh glow of the morning. I felt as if all my dark past—the wrestling and struggling with poverty and inadequacy that had wasted me for ages and ages, had dropped away with the passing of night and I found myself born again into a new heaven and a new earth" (Fisher Papers, n.d.). However, in the months that followed, Yezierska found it increasingly difficult to adjust to her new environs and decided to return to New York City. Canfield graciously expressed regret when she learned of Yezierska's decision, but made no effort to change it. In a letter entitled "Wednesday evening," Canfield cautioned Yezierska to ensure that the "dark and dreary weather" had not unduly influenced her decision to leave Vermont (Fisher Papers, n. d.). However, in a later note appended to the correspondence between the two, the native identified Arlington's chilly social climate as the primary reason the uprooted writer had not transplanted successfully:

> She was . . . a Russian Jewess, of little education and a very emotional temperament who had written a few intense books about (I think) Jewish immigrants in New York out of her own experience. . . . Her efforts to get in touch with the Vermonters of Arlington, their efforts to understand her enough to help her—all proved futile. The

> enormous psychological differences between them were impassable. . . . Neither could understand a single impulse or thought or emotion of the other. She . . . weeping . . . distraught, went away after having lived here, I think, more than a year. (Fisher Papers)

This note reveals a matter-of-fact formality that proves the limitations of Canfield's reputation as Vermont's resident welcome wagon, for it shows that she considered Yezierska as a sort of strange fruit, a peach or pomegranate tree that could not survive in an apple orchard.[12] As opposed to the persona Canfield had cultivated in her fictional invitations and in brochure advertisements, Yezierska learned a discomfiting truth about the woman celebrated as the "First Lady of Vermont." Contrary to popular belief, the doors to the writer's home and her heart could slam shut just as quickly as they opened.

An analysis of the connections between Canfield's writings about Vermont tourism and the Vermont Eugenics Survey would not be complete without an evaluation of *Seasoned Timber*, the 1939 novel that Canfield considered her masterpiece. Her contemporaneous readers apparently shared this sentiment, since the writer received more fan mail about this novel than any other book.[13] Reviewers agreed that *Seasoned Timber* was the novel in which Canfield's progressive social attitudes and her commitment to the "durable Vermont traditions" of integrity, democracy, and social equality were most fervently stated (Review 1939). But in addition to these ideals, the plot of the book eerily echoes the reservations about tourism and foreigners that surface in her earlier writings. The story depicts the personal and professional crises of Timothy Canfield Hulme, the headmaster of a private secondary school in Clifford, Vermont, that is in danger of closing due to financial difficulties. A wealthy trustee, Mr. Wheaton, has offered to endow the academy, but only if Hulme agrees to deny admittance to Jewish students. The offer disgusts Hulme, but he is unsure if he can convince the townspeople to reject it.

Much of the action in the novel depicts the distraught headmaster's attempts to convince the citizens of Clifford to reject Mr. Wheaton's proposal. At the climactic town meeting, Hulme threatens to resign if the town votes to accept the money. His job is saved after the townspeople, following the historic example set by the "Green Mountain Boys," make the village safe for democracy by siding with Hulme (1939, 357). However, Canfield's lengthy novel does not end here; it continues with a denouement that casts quite a different light on the headmaster's morals. After news of the town's decision spreads nationwide, the academy is deluged with applications and Hulme graciously invites the parents of prospective students to "look the Academy over." The headmaster later confides to a friend that such visits will give him a chance to look *them* over in order to determine if the students who "sound desirable" can pay their tuition bills (440). Despite the headmaster's self-congratulatory sense of honor (metaphorized as a tree in the title of the novel), this scene, as well as a subsequent chapter in which Hulme denies admission to a Jewish boy but then admits him so that the student may "escape"(145) from his "awful Jewish mother" (307), demonstrates that Hulme's character, like the tree Brewer employed to symbolize New England's stock, is rotten inside.[14]

These episodes, as well as the novel's conclusion, reflect the sinister strains of non-accommodationist thinking that pervaded Canfield's earlier writings about tourism. As a result of an "old home" vacation that includes exposure to Hulme's leadership during the Wheaton conflict and rediscovery of his family's roots, the headmaster's nephew Canby decides to settle permanently in Clifford and make it possible for others to do so by renovating abandoned houses into summer homes. While describing the plan to his uncle, Canby reveals that his motivations are not merely economic, but eugenic; he intends to sell houses only to "educated people" whose presence will uplift "these little old back towns" before, as his wife Susan emphasizes, it is "too late" (477). In the final pages of the book, as a symbolic affirmation of Canby's determination to raise the region's stock, Susan gives birth to a baby boy. The new family thus represents the successful fruition of the reproductive agendas articulated by the Vermont Eugenics Survey, the Vermont Commission on Country Life, the Vermont tourist industry, and most significantly, by Dorothy Canfield herself over the course of twenty-four years.

To examine Dorothy Canfield's writing in the context of eugenics is to encounter an interpretive paradox. Though in her published and private writings Canfield frequently and vehemently excoriated incidents of anti-Semitism, racism, and ethnic prejudice, her works advocating Vermont tourism express discriminatory sentiments about the dangers the "wrong" kinds of people posed to the state's carefully constructed reputation as a vacationer's paradise. While the Canfield writings I have discussed portray Vermont as the locus of America's characteristic virtues, including patriotism, democracy, and equal opportunity, these texts also demonstrate the eugenic agenda that state leaders, scientists, and citizens including herself had embraced. It is fascinating, and indeed troubling, that although a blatant undercurrent of social discrimination is clearly evident in her writing, she never acknowledged the contradictory messages it carried. The same writer, who empowered Timothy Hulme to fight Wheaton's proposal with the words "Now is the time . . . for all good men to stand up for their country"(319), used the same rallying cry at the end of a speech entitled "How We Can Keep On Being Vermonters." In this speech she exhorted her audience to preserve the "Vermont qualities of decency, simplicity, and admiration for fine character" by recruiting the kind of summer folks "we like and enjoy having, the kind that would do us good, as well as help us make a living."[15] The virtual contradictions apparent in these symbolic calls-to-arms indicate that Canfield imagined Vermont not only as a geographic location, but also as a state of mind that citizens like herself had the patriotic duty to defend. That eugenics emerges as a constitutive element in the work of such an esteemed social activist indicates the extent to which this disturbing "science" captured the imaginations not only of nativist researchers here and the Nazi regime abroad, but also of model citizens at home—a fact we would do well to remember when analyzing the impact of eugenics on American literature and culture.

Notes

I would like to thank Sarah Tracy and Elizabeth Fowler Johnson for their comments on this essay.

1. See for example Joan Shelley Rubin 1992, Ida H. Washington 1982, and the work of Mark Madigan.
2. A note on nomenclature is necessary here. For income tax purposes, Fisher published all of her fiction under her maiden name. She used her full married name as the byline on nonfiction pieces and in Book-of-the-Month Club contexts. To avoid confusion, I refer to the writer as "Canfield" since the majority of her works that I discuss herein are fictional.
3. Letter from Dorothy Canfield to "Grandfather," June 29, 1888. Eli Hawley Canfield papers.
4. See especially the stories "A Fishing Excursion" and "The Village King." Fisher papers.
5. Letter to Alfred Harcourt, 4 Oct. 1911. Fisher papers.
6. In her article on "Dorothy Canfield Fisher's *Tourists Accommodated* and Her Other Promotions of Vermont," Ida H. Washington does not associate the stories in *Hillsboro People* with tourism. Rather, she maintains that the book

exemplifies "basic human problems as they occur in the Vermont setting" (1997, 162).

7. Canfield repeats this description of a French Canadian character in her Hillsboro stories (1915, 282, 287, 292) and in her 1921 novel *The Brimming Cup*, in which Neale Crittenden refers to the employees at his lumber mill as "hot-tempered irresponsible Canucks" (201).

8. Though the pedigree charts Perkins compiled established his findings as irrefutable scientific facts (Dann 1991, 12), the criteria he used to judge familial "degeneracy" were decidedly subjective. He included among the ranks of Vermont "defectives" individuals who were "illegitimate," and "insane," as well as people who were inexplicably "a little odd" or "not just right" (11).

9. Letter from "Harry" Perkins to Guy Bailey, 16 November 1928; subsequent undated letter in Guy Bailey papers; letter from Guy Bailey to Harry Perkins, n.d. 1932, Guy Bailey papers.

10. It is interesting to compare the report of Canfield's committee with the recommendations of the Commission's Committee on Summer Residents and Tourists. The latter group believed that "refined" automobile tourists would positively benefit the state because in addition to financial profits they would give rural children an opportunity to meet "cultured" people (Two Hundred Vermonters 1931, 124).

11. Letter from Yezierska to Dorothy Canfield Fisher, "Thursday," n.d. Fisher papers.

12. Here I appropriate a New England proverb that Canfield often quoted: "Peaches and pomegranates do not, you see, grow on apple trees. But apples do" (Canfield Fisher 1953, 392). Although Yezierska effusively thanked Canfield for her friendship in later correspondence, she was able to read between the lines of her friend's letter and later denounced her hypocritical behavior in *Red Ribbon on a White Horse*.

13. Letters to Alfred Harcourt, 4 December 1938 and 20 November 1939. Fisher papers.

14. The appearance of the "Jewish mother" in *Seasoned Timber*, as well as her predecessor in *Tourists Accommodated*, reflect the anti-Semitic tenor of Vermont tourism in the first part of the twentieth century. Artifacts such as hotel brochures and advertisements for lodging prove that many tourist establishments actively discriminated against Jews (Calder 1997).

15. Typescript of "How We Can Keep on Being Vermonters" (address to the Vermont Federation of Women's Clubs, n.d.), 6. Fisher papers.

References

Bailey, Guy. Papers. University of Vermont.

Brewer, Daniel Chauncey. 1926. *The Conquest of New England by the Immigrant*. New York: G. P. Putnam's Sons.

Brown, Dona. 1995. *Inventing New England: Regional Tourism in the Nineteenth Century*. Washington, D.C.: Smithsonian Institution Press.

Calder, Jacqueline, curator. 1997. Exhibit entitled *Tourists Accommodated: Visiting Vermont, 1895–1995*. Vermont Historical Society.

Canfield, Dorothy. 1915. *Hillsboro People*. New York: Henry Holt and Co.

———. 1921. *The Brimming Cup*. New York: Harcourt, Brace and Co.

———. 1932. *Tourists Accommodated.* New York: Harcourt, Brace and Co.

———. 1939. *Seasoned Timber*. New York: Harcourt, Brace and Co.

———. "How We Can Keep on Being Vermonters" (typescript). Dorothy Canfield Papers. Bailey-Howe Library, University of Vermont.

Canfield, Eli Hawley. Papers. Vermont Historical Society.

Crane, Charles Edward. 1937. *Let Me Show You Vermont.* New York: Knopf.

Dann, Kevin. 1991. "From Degeneration to Regeneration: The Eugenics Survey of Vermont." *Vermont History* 59, 5–29.

Fisher, Dorothy Canfield. 1949. "Vermont Summer Homes." *Vermont Life* (Spring): 3–7, 60.

———. 1953. *Vermont Tradition: A Biography of an Outlook on Life*. Boston, Mass.: Little, Brown.

Fisher, Dorothy Canfield. Fisher Papers. Bailey-Howe Library, University of Vermont.

Goldman, Hal. 1997. "A Desirable Class of People: The Leadership of the Green Mountain Club And Social Exclusivity, 1920–1936. *Vermont History* 65 (Summer): 131–52.

Madigan, Mark J. 1996a. Introduction. *The Bedquilt and Other Stories by Dorothy Canfield Fisher*. Columbia: University of Missouri Press.

———. 1996b. Introduction. *Seasoned Timber*. By Dorothy Canfield. Hanover, N.H.: University Press of New England.

Perkins, Henry F. 1938. "Hereditary Factors in Rural Communities." *Eugenics* 3, unpaginated.

Rebek, Andrea. 1997. "The Selling of Vermont: From Agriculture to Tourism." *Vermont History* 65, 14–27.

Review of *Seasoned Timber*. 1939. *Burlington Free Press* (28 March). Fisher Papers. University of Vermont.

Rubin, Joan Shelly. 1992. *The Making of Middlebrow Culture*. Chapel Hill: University of North Carolina Press.

Two Hundred Vermonters. 1931. *Rural Vermont: A Program for the Future*. Burlington: The Vermont Commission on Country Life.

Washington, Ida H. 1982. *Dorothy Canfield Fisher: A Biography*. Shelburne, Vt.: New England Press.

———. 1997. "Dorothy Canfield Fisher's *Tourists Accommodated* and Her Other Promotions of Vermont." *Vermont History* 65, 153–64.

Yezierska, Anzia. 1987. *Red Ribbon on a White Horse: My Story*. London: Virago.

Eugenics and the Experimental Breeding Ground of Susan Glaspell's *The Verge*

Tamsen Wolff

SUSAN GLASPELL'S 1921 PLAY *THE VERGE* CONTAINS A MOTLEY COMBINATION of styles, genres, and influences; faced with this mix, contemporary reviewers of the production consistently noted Glaspell's scientific allusions, particularly to the work of Luther Burbank and Hugo de Vries.[1] De Vries (1848–1935), a Dutch botanist, advanced the theory of hereditary mutations and was credited with being one of three scientists who rediscovered Gregor Mendel's laws of heredity in 1900[2] ; while Luther Burbank (1849–1926), a horticulturist from Massachusetts, created hundreds of new plant varieties by hybridization, including the Shasta daisy and the hardy Burbank potato. For the theater critics of Glaspell's play, the different accomplishments of these scientists were easily grouped together because both drew on a still thrillingly new concept of heredity to demonstrate the transformation of species.

Glaspell's novels and plays frequently investigate questions of inheritance, mutation, transmission, and descent, but none engages more imaginatively with the question of what is biologically and socially transmissible than *The Verge*.[3] The overarching event, and symbolic heart of the play, is the parallel experiment in horticultural and human mutation carried out by the play's heroine, Claire Archer. Claire, an upper-middle-class woman approaching middle age and in her second marriage, has recently become consumed with botanical experimentation and the possibility of creating whole new species. Feeling trapped by her Puritan, Anglo-Saxon heritage, Claire is equally captivated by the possibility of transforming her own identity. In addressing the issue of hereditary experimentation with plant and human life, Glaspell confronts as well the ideas and rhetoric of the eugenics movement. What is more, eugenics proves provocative for Glaspell and informs the theme, language, characters, design, and structure of *The Verge*.

The American eugenics movement was dedicated to "the improvement of the human race through better breeding" (Davenport 1911, 1). Eugenics and its vocabulary were pervasive; the word "eugenics" was heard from infancy in America for the first three decades of the twentieth century (Hasian 1996, 30).[4] Eugenics was predicated on Mendelian hereditary theory and was responsible for disseminating what promptly became accepted conceptions about that theory. Throughout the nineteenth century, various speculative theories about heredity were floated, almost all of them combining theories of evolution and heredity. The first and most striking claim of Mendelian theory was the recognition of distinct hereditary units, a theoretical move that separated reproduction from growth, focusing exclusively on reproduction and the transmission of hereditary material. In 1865, Mendel had demonstrated in experiments with edible garden peas that hereditary material is transmitted intact from parent to offspring.[5] Mendel determined that heredity is governed by discrete hereditary "factors" or "elements" (later "genes"), which maintain their integrity and are not altered when combined. Consequently, Mendel, posthumously and inadvertently, provided the foundation of the popular eugenic view that internal factors functioned completely independently of the external environment, and thus, biological inheritance could neither be resisted nor diverted.[6] The resurrection of Mendel's theory in 1900 was galvanizing to its early proponents for several reasons: it was predictive, generalized, experimentally defensible, and applicable to all living organisms, including humans. The years 1900 to 1926 saw a flurry of developments, counterdevelopments, and refinements to Mendel's theory, which would culminate in Thomas Hunt Morgan's *The Theory of the Gene* in 1926. Eugenicists, however, promoted the earliest and simplest version of Mendelism in order to support a range of overlapping goals, from professional advancement to social control.

Glaspell could not have addressed explicit questions of heredity at this time without taking on eugenics, since issues of hereditary experimentation did not divide neatly into distinct eugenic and genetic categories.[7] If anything distinguished eugenics from other emerging disciplines investigating and using hereditary theory (including cytology, genetics, and evolutionary studies), it was eugenicists' particular set of convictions: namely, that Mendelian concepts are inescapable and universally applicable; that they are applicable to every human characteristic—physical, mental, emotional, moral, occupational, and behavioral; that applying hereditary theory to human beings is both a logical and a moral imperative, as well as a socially necessary step; and that doing so will carry the human race towards a narrowly defined ideal of beauty and purity. In *The Verge*, Glaspell engages with many of the contentions of the eugenics movement and with the value-laden language that enforces essential points.

Glaspell grapples with one concern especially that is deeply entrenched in the discourses of eugenics: the possibilities of autonomous direction and mobility. Eugenicists frequently assert the conviction that heredity is predetermined and fixed, which implies the impossibility of individual self-formation, development, or adaptive control. G. K. Chesterton, in his astute and idiosyncratic analysis of eugenics, *Eugenics and Other Evils*, claims that eugenic theory negates free will. For the eugenicist "master," an individual is "*moveable.* His locomotion [is] not his own: his master move[s] his arms and legs for him as if he were a marionette" (1987, 362). For Glaspell, this preoccupation could be summed up in the question: how much can anyone determine who he or she is? The struggle to do so entirely is personified by Claire and her need to escape "the forms molded for us," but the struggle is not hers alone (1. 64).[8] All of the characters are constrained (and sustained) by inheritance, and they demonstrate a continuum of desire and ability to wriggle out from under biologically or socially prescribed roles.

Assumptions about the omnipotent gene led eugenicists to posit the end of genetic uncertainty or speculation; as one poster at the 1920 Kansas Free Fair proclaimed, "What you really are was settled when your parents were born" (Fairs 1920–1925). Although here biological determinism and the genetic past override any possibility of individual direction, at the same time, the eugenic emphasis on transmission does not simply remove individual agency. Darwinian natural selection emphasizes the arbitrariness of an individual's happening to be in accord with the demands of his or her environment and, as a result, of his or her survival. By way of contrast, eugenics at least appears to place the power to make decisions about heredity within an individual's grasp. In eugenic propaganda, people are the unavoidable product of heredity and the past, but they also possess the ability to take charge of the next generations. As another banner at the Kansas Free Fair declared, "the future of the race is up to you" (Fairs 1920–1925).

In negotiating the place of individual autonomy and direction within hereditary theory, Glaspell also had to take into account eugenicists' assertions of progress. In contrast to Charles Darwin (and other evolutionists after him), who had argued against the equation of evolution with progress, eugenicists jubilantly coupled hereditary theory with progress. Following Darwin's reasoning in *The Origin of Species*, evolutionists also maintained that within each species variation, rather than truth to type, is the governing principle of evolutionary development. But for eugenicists breeding humans for a selective, specific, "better," type—and thereafter for truth to that type—constitutes the mission of the movement. The faithful attendants to the marriage of biological determinism and progress then are the joint premises of purity and beauty. Eugenicists' constant harping on these themes means that some of the vocabulary used to describe or promote hereditary theory, which we would not now recognize as noteworthy, was charged with controversy and new associations. With the rise of eugenics, certain adjectives particularly were used relentlessly to describe the pragmatic results of the eugenic project, including "better," "beautiful," "pure," "sane," "healthy," and "happy"; these six words collectively appear over seventy times in one chapter of the popular eugenic primer, *The Fruit of the Family Tree* (Wiggam 1924). Glaspell's use of this vocabulary in *The Verge* is part of what reveals her simultaneous urges: to disavow fixed identity and the eugenic goals of a specific human type and also to embrace the developments in hereditary theory that allow for the possibilities of new forms as a result of experimental breeding.

The opening action of *The Verge* takes place in winter in a greenhouse that has drawn the heat from the main house, at Claire's request, for the sake of the plants. After her initial offstage action—a phone call to alert the sympathetic gardener, Anthony, to what she has done—the members of the indoor household are out in the cold. They have to relocate from the chilly domestic scene to the heated and germinating greenhouse of their unusual hostess. This forced migration from a familiar home ground to a strange site of experimentation immediately focuses attention on the question of how space affects meaning. Much of the initial dialogue and action is about what and who belong in this space: the smell of fertilizer "belongs here," whereas Claire's husband, Harry, and their guests—her lover, Dick, and her confidant, Tom—do not (1. 61). As the play's title indicates, Glaspell is concerned with boundaries: between sanity and insanity, belonging and otherness, plant and human, fixity and change, old and new, pattern and creation. In her biography of her husband, Jig Cook, Glaspell devotes an entire chapter to a description of a greenhouse that he built and about which he composed a poem. In the poem, Cook's "amazing Greenhouse . . . celebrates its walls, rampart against a thousand leagues of cold invading from the bitter polar night"; here the greenhouse walls are heralded for what they keep out (Glaspell 1927, 201). However, the glass walls in *The Verge* are also very much about what is let in: light, sound, action, and, when Claire chooses to unlock the door, the other characters. The solid yet permeable greenhouse walls provide an indoor environment that allows for aspects of the outside world to infiltrate, highlighting the inside/outside dichotomy and the continuum between the two.

Before the descent of Claire's husband and guests, the audience has already encountered the greenhouse, which is dominated by plant life that Glaspell describes in detail:

> The Curtain lifts on a place that is dark, save for a shaft of light from below which comes up through an open trap-door in the floor. This slants up and strikes the long leaves and the huge brilliant blossom of a strange plant whose twisted stem projects from right front. Nothing is seen except this plant and its shadow. . . . [The lights come up and at] the back grows a strange vine. It is arresting rather than beautiful. It creeps along the low wall, and one branch gets a little way up the glass. You might see the form of a cross in it, if you happened to think that way. The leaves of this vine are not the form that leaves have been. They are at once repellent and significant.[9] (1. 58)

This Edge Vine, so named because it was poised on the edge of mutation (although now is "turning back" and not taking the leap), is Claire's most recent experiment (1. 61). The next highly anticipated hybrid, called Breath of Life, is due to flower in forty-eight hours, or at the end of the play.

The imaginative visual power of the plant is complemented by the pragmatic purpose of this space. The stage directions indicate that the greenhouse is not about decorative displays or mundane labor, but is instead a serious "place for experiment with plants, a laboratory" (1. 58). Glaspell conceives of the space itself as active, in the same way that Cook had incited his greenhouse:

> No mere Wordsworthian guest of nature be,
> Spectator and not sharer of her life,
> But her co-worker, with selective art
> Prescribing form to her wild energies:
> Saying, 'Thou shalt be!' and 'Thou shalt not be!' (Glaspell 1927, 202)

The greenhouse Cook personified (and Glaspell then immortalized for him) is not simply a place in which someone else experiments, but one that will contribute to the experimentation itself. In her dramatic greenhouse, Glaspell similarly attempts to emphasize the agency of the site.

At the start of *The Verge*, the audience is looking at human beings, some engaged in experiment, who are themselves under glass, or in a teeming, theatrical petri dish. The greenhouse is a place for disrupting nature, and the space is itself disrupted by the comings and goings of unpredictable human elements. Claire aligns herself immediately with the experimental plant life. When Harry calls her "the flower of New England," Claire wants only to "get away" from her ancestors and to attempt for herself what she attempts with the plants: "to break it up! If it were all in pieces, we'd be shocked to aliveness—wouldn't we? There'd be strange new comings together, mad new comings together" (1. 64). If the plants are placed in artificial circumstances in order to create different crossings, the characters are revealed newly in their couplings in part as a result of being transplanted to this foreign ground. When Tom and Dick are alone for a moment in the greenhouse, for instance, Tom suddenly notices himself in relationship to Dick: "I had an odd feeling that you and I sat here once before, long ago, and that we were plants. And you were a beautiful plant, and I—I was a very ugly plant" (1. 73). His vision of his own ugliness signals his consequence, since it corresponds to the "repellent and significant" countenance of the Edge Vine, as does his surname, Edgeworthy. Furthermore, his musings encourage the audience to continue completing the equation of plants and humans, and to draw their own conclusions about what those counterparts might look like or might mean.

Into this theatrical site, animated reciprocally by design and audience interpretation, Glaspell places Claire in the deliberately active, conscious role of an experimenter, who sets out to disturb and disrupt the conditions that she observes. The distinction between observer, laborer, and experimenter is important to Glaspell. French physiologist Claude Bernard's definition holds for Glaspell: "We give the name experimenter to the man who applies methods of investigation, whether simple or complex, so as to make natural phenomena vary, or so as to alter them with some purpose or other, present themselves in circumstances or conditions in which nature does not show them" (1949, 15). Following this model, Claire takes her experimentation seriously, striving to maintain constant controls in her laboratory, documenting developments, and repeating experiments with tiny adjustments.

Yet Glaspell expands on the definition of controlled experiment because Claire seeks to experiment, with precision, in order to achieve something fundamentally *unfixed*. Claire's vision, and constant refrain, is of a mutated hybrid that will continue "alive in its otherness" and never harden completely into a set form (1. 62). Interestingly, several critics accuse Claire of being an "unconvincing" or "fraudulent" character because she practices what they take to be bad science, genetic experimentation without a declared, useful goal (Parker 1921, 296, "Claire" 1921, 4). As Robert Parker notes with irritation in *The Independent*, "The authentic scientist today does not indulge in botanical hocus-pocus and melodramatic mutations" (1921, 296). Of the two "authentic" contemporary or near-contemporary scientists whom critics assumed were models for Glaspell, Hugo de Vries focused specifically on mutation in the American evening primrose, suggesting that evolution proceeded as a result of sudden radical changes in the characteristics of varieties. Claire expresses de Vries' conclusion this way: "Plants do it. The big leap—it's called. Explode their species—because something in them knows they've gone as far as they can go" (1. 71). At the same time, Claire's aim of creating scent for the hybrid flower Breath of Life, a fragrance she calls "Reminiscence. Reminiscent of the rose, the violet, arbutus—but a new thing—itself" (1. 64), concurs with Luther Burbank's attempt to create a variety of calla lily with a layered scent, a flower that he called "Fragrance." Burbank explained his wide-reaching experiments in hybridization, which lasted over fifty years, by saying modestly, "I shall be contented if because of me there shall be better fruits and fairer flowers" (Beeson 1927, 14). If his experimentation caused what he termed "perturbation" in the plants, this disruption was in service to his goal of betterment resulting from wider variation. What makes Claire's botany "hocus-pocus," and her mutations "melodramatic," has to do with her resistance to the eugenic frenzy around the concepts of beauty, usefulness, and advancement that Burbank, along with many others, invokes.

Harry comments on Claire's refusal of these ideas almost immediately in the play, and later her resistance is elaborated in an important exchange with her seventeen-year-old daughter, Elizabeth, who is visiting from boarding school. In a conversation with Dick over breakfast, Harry hankers after the time when Claire made "flowers as good as possible of their kind," before she started going around the bend, "making queer new things" (1. 64). At the end of the act, Elizabeth attempts to make sense out of this development by attributing the eugenic viewpoint to her mother's radical experimentation: it is Claire's "splendid heritage" that gives her the "impulse to do a beautiful thing for the race" (1. 75). The "beautiful thing" immediately apparent to Elizabeth is Claire's wish "to produce a new and better kind of plant" (1. 76). This Claire forcefully rejects, saying, "They may be new. I don't give a damn whether they're better. . . . [As if choked out of her] They're different" (1. 76). Six times Elizabeth presses the same point ("What's the use of making them different if they're not better?" [1. 76]), and each time, Claire counters furiously, deeply troubled by the eugenic reasoning, and especially by the ominously freighted words "useful," "beautiful," and "better." The fraught misunderstanding between mother and daughter escalates until Claire violently uproots and kills her creation, the Edge Vine—a metaphorical substitute for Elizabeth—for reverting to conventional type. Elizabeth drives home the point for Claire that an experiment—whether a daughter, or a plant—can fail and not be redeemable.

Claire occupies an unusual position as a kind of non-mother: she disowns her living daughter, and her only other child, a four-year-old son David, has died within the past two years. Claire is as unhappy about David's absence as she is about Elizabeth's vacuous and rigid presence. In other words, Claire does not renounce a large, abstract concept of "motherhood," like the one painted in broad strokes by eugenicists, who create a "sweeping silver-tongued rhetoric about pure motherhood" (Chesterton 1987, 297). Instead, Claire is at odds with one specific child, her daughter. Yet she predicts that she might have had a problem with her son too, had he lived. As she explains it, she loved her son because he displayed mobility. She tells Tom, "I always loved him. He was movement—and wonder. In his short life were many flights" (2. 87). Still, afraid that David, like herself, "would always have tried to move and too much would hold him," Claire claims to be "glad" of his death (2. 87). Having lost one child, and refused to raise another, Claire does not derive her identity from her genetic role. At the same time, through Claire's poignant reminiscences about her son, Glaspell acknowledges the loss in Claire's disavowal of maternity.[10]

In Claire, Glaspell continually points out the forced, artificial opposition of biological *reproduction*—the designated female province—and *production and self-production*—a conventionally male domain. Glaspell attempts to blur the line between the two, viewing hereditary experimentation, and mutation especially, as a process that encompasses both reproduction and self-production. The innocuous, traditionally feminine project of raising flowers is, as Harry observes, "an awfully nice thing for a woman to do," until Claire turns it into something threatening, something, as Harry says, "unsettling for a woman" (1. 65). When Claire refuses reproduction and insists instead on self-production, for plants or herself, she cuts herself off from easily identifiable female behavior. Isolated by her desire, Claire toys increasingly with the prospect of insanity to provide the "otherness" she seeks, even as she struggles to communicate with other people.

In the scene with Elizabeth, Claire has more and more difficulty speaking, which introduces Glaspell's great concern with the role of language and speech in reconceiving or thwarting heredity. By the second act, the use or misuse of words replaces the use or misuse of plants as the main subject. To emphasize the difficulty of navigating words, Glaspell moves the action to a small, enclosed tower, where Claire is backed into an increasingly narrow physical and verbal corner.[11] Reportedly, Margaret Wycherly, who played Claire in the original production of the Provincetown Players' 1920–21 season, asked Glaspell to write a play with a vocal female character; Glaspell sent her the manuscript of *The Verge* describing it as "a play in which a woman speaks" (Provincetown 1921). Claire does speak, but it is her conscious struggle to articulate, and her desire to live in that struggle, that set her apart. As Ludwig Lewisohn, perhaps Glaspell's most insightful contemporary critic, describes the playwright: "She is a dramatist a little afraid of speech" (1932, 393).

The traps of language that Glaspell wants to expose and avoid include the expansive imprecision typical of eugenic vocabulary and the way in which eugenic syntax is devoid of agency. Together, these characteristics produce the forebodingly unidentified, scientifically legitimized, universalizing, and morally weighted authority of eugenic rhetoric. As G. K. Chesterton asks, "Who is the lost subject that governs the Eugenist's verb? [And] . . . What is this flying and evanescent authority that vanishes wherever we seek to fix it?" (1987, 50). When Claire's sister, Adelaide, visits Claire in the second act, her speeches are suffused with a similarly elusive authority about the broad social and biological forces that necessarily determine identity and behavior. At the same time, Adelaide, Harry, and Dr. Emmons, a nerve specialist summoned to examine Claire, all rely on limited definitions to delineate for Claire "the woman you were meant to be" (2. 79).

Limiting terminology is a staple of eugenic literature. The eugenic doctrine depends on the equivalency—extrapolated from Mendel—of peas to people, which Glaspell parallels in her conflation of plants and people. The jump from peas to people is a slippery rhetorical elision that enables eugenicists to classify an individual person as an object, a number (used by eugenicists to identify the hereditary traits of an individual), or a commodity (e.g., a "high-grade" defective).[12] The obvious, limited, symbolic world constructed by eugenicists aims to show in no uncertain terms the thingness, the typology, of its subjects. Moreover, a generalized refusal to distinguish subject and object means that forms of self-making become inseparable from a radical removal of selfhood, in which an individual is identified only in terms of a collective or a representative type.

Glaspell's creative response to the linguistic limits and pitfalls characteristic of eugenics takes several forms. First, she expands on and explores the eugenic reduction of people to types. Theater theorist, director, and designer Edward Gordon Craig, with whose work Glaspell was familiar, was interested in the possibility of symbols registering inner reality; eugenicists, in contrast, used symbols to flatten, or negate, subjects' inner realities. Nonetheless, the eugenic sensibility of representative specimens is suggestive of both expressionism and symbolism. In *The Verge*, the characters' names are emblematic.[13] Claire Archer aims at clarity, while Tom's name, Edgeworthy, suggests his readiness at least to perch on the edge of change, if not to take the plunge. Yet, both of their designations point up the attempt to exceed type, rather than to embrace or even to adopt it. Playing further with stereotyping, Glaspell creates the proverbial triumvirate of Tom, Dick, and Harry, and even establishes a hierarchy among the three—soulmate, lover, husband—by naming them in descending order of their comprehension of Claire and her objectives.

In presenting and counteracting the problems with the language surrounding hereditary theory, Glaspell also suggests alternative languages. First, she shows the possibility of a nonliteral, gestural language, most plainly in the first act when the characters are attempting to communicate with one another through the glass wall of the greenhouse. Claire wins the competing dance of communication hands down when her suggestive action—a sneeze—is immediately understood by Tom to mean that the greenhouse inmates need pepper, while Harry's noisy, realistic reenactment of his need for salt leaves Tom completely perplexed. Claire's propensity to gesture instead of speak continues to demonstrate the potential usefulness of a language of the body rather than one mainly or solely of the voice. Moreover, when Claire does speak, Glaspell tries to address the collapse of signifier and signified, what Claire calls "smothering it with the word for it" (2. 91). In order to keep the gaps between meaning and words alive, Glaspell constantly wields her syntactical ally, the dash. In part by so doing, Glaspell suggests that we are not best served by precision in language, any more than we are by the deceptively passive, wide net of eugenic rhetoric. Finally, Glaspell illustrates the value of a visual language in the play's design, in its emphasis on expressionism, particularly in the evocative tower of the second act, and symbolism, in the weighted plant imagery of the bookend acts. The designs of *The Verge* pose multiple interpretive possibilities.

While the forced limits of eugenic discourse are consistently suggestive to Glaspell, the richness of its contradictorily broad, necessarily creative argument is as well. Although the premise of the eugenic theoretical project is, as Elizabeth so clearly stated, the purification and elevation of the human race, the theory itself is a composite of contradictions (including, for example, the dangers of hybridity and the dangers of inbreeding). Eugenics is not simply wrong-headed theory, it is often enthusiastically, imaginatively, wildly wrong-headed. Luther Burbank's claim that "stored within heredity are all joys, sorrows, loves, hates, music, art, temples, palaces, pyramids, hovels, kings, queens, paupers, bards, prophets and philosophers . . . and all the mysteries of the universe" is characteristic, if relatively benign, eugenic hyperbole (1907, 83). In an effort to show how human beings' social existences and physical and psychological attributes are governed by scientific determinism, eugenics draws heavily on a multitude of analogies, equivalencies, images, and confirming metaphors for causation. As eugenic theory became less and less scientifically sustainable, the fictive leaps necessary to maintain its position as "science" became bigger and bigger, and the analogies eugenicists used to cover these gaps in logic became increasingly elaborate. Thus, eugenics manages to be an overdetermined, inflexible theory that is also in a generative, almost frantically fictive state.

Glaspell's fascination with the possibilities and limitations of language, with the power implicit in naming and creating, mirrors the heated contemporary scientific debate about how to name or describe that brand new and invisible entity, the gene. Proponents of a specific and original vocabulary included Danish scientist, W. L. Johannsen, who wrote in 1911, "It is desirable to create a new terminology in all cases where new or revised conceptions are being developed. Old terms are mostly compromised by their application in antiquated or erroneous theories and systems, from which they carry splinters of inadequate ideas, not always harmless to the developing insight. Therefore I have proposed the terms 'gene' and 'genotype'" (cited in Darden 1991, 182). Glaspell illustrates a similar desire in Claire to cut loose from "antiquated or erroneous theories and systems" in *The Verge*, but Glaspell recognizes as well the impossibility of the clean break that Johannsen seems to want. When Claire labors to achieve a freedom of expression with Tom at the end of Act II, she can only accomplish a kind of transcendence by reverting to the formal structure of verse. Her attempt at a new form of language is shaped by the reemergence of an old one.

From the start of the second act, Glaspell focuses less on how to escape the past and more on how bits of old forms, feelings, and definitions materialize in and affect the present. Mixed up with the glorified big leap to the new is a simultaneous longing for aspects of the past, a need for what Claire calls "the haunting beauty from the life we've left" (2. 86). One emissary from that past is Christianity, surfacing in the play in the form of a hymn, *Nearer My God to Thee*, which serves as a leitmotif for Claire.[14] The beloved "ancestor's hymn" is nonetheless troubling to Claire because the old religion is contending with a new religion of scientifically controlled breeding. In charting where and how the past asserts itself in the present, or the present displaces the past, Glaspell is especially attentive to what is lost in the transition. Claire's efforts to transcend set ideological positions extract a price: her rejection of a maternal position leaves her bereft, and her conflict over whether to conduct a romantic affair with Tom leads to an even greater loss. In the greenhouse in the final act, Breath of Life achieves its mutation, but Claire's corresponding struggle to remain actively in a state of otherness results in her choking Tom to death when he attempts to bring her back to sanity and romantic love. The implausibly easy and rapid murder underscores Glaspell's suggestion that sacrifice is a part of transformation. In her altered state, nearly inarticulate, Claire continues to be buoyed by the religious past as she slowly, painfully sings *Nearer My God to Thee* while the curtain falls. Glaspell's constant attention to the ways in which shreds of ideas and past forms cling to and inform present developments complicates the multiple, overlapping theatrical languages in *The Verge*.

Since the action of the stage necessarily occurs in the present, dramatists consistently confront the problem of how to display the past on stage. For a modern dramatist like Glaspell, who focused on developing complex relationships between the past and the present on stage, eugenic insistence that heredity is visible in the embodied present was especially suggestive. In addition, the kind of creativity that was necessary to keep eugenics afloat, combined with its topicality, made its ideas and vocabulary fertile ground for the imaginations of writers, and of playwrights in particular, for theater lends itself to analogy through visuals, as Glaspell demonstrates in her ingenious use of plants. Moreover, the eugenic project itself is a similar enterprise to the project of modern drama. The inchoate study of genes parallels the modern dramatist's exercise of imagination. Both the scientist and the dramatist are attempting to hazard, legitimate, and defend that which has not yet been seen: a new form, either that of a gene (and a corresponding professional movement), or that of a play (and a modern American theater).

Venturing these new forms results in messy compositions, and *The Verge* is a play in which the seams show. Multiple forms—melodrama, comedy, expressionism, and realism—repeatedly jostle against one another. Glaspell makes use too of a wide range of thematic sources, and while the separate strains are worth individual attention, they work in concert, without priority, to intensify and complicate an overarching contemplation of heredity.[15] The very confluence of new and old forms and influences reflects Glaspell's aim to create a new dramatic hybrid, unfixed in its form yet rooted in a charged, contemporary debate. *The Verge* is Glaspell's most original play, and eugenics offered an ideal frame of reference to help make that invention possible.

Notes

1. Critics who cite Glaspell's scientific sources include Young (1921, 47), Macgowan (1921, n.p.), Rathburn (1921, 4), Hale (1921, sec. 6, p. 1), and Parker (1921, 296).

2. Critics have challenged the mythologized, three-way rediscovery of Mendel's ideas. See, for example, Bowler 1989.

3. The complexities of inheritance—of convention or blood—are addressed in Glaspell's other 1921 play, *Inheritors* (which contains a more literal engagement with some of the same questions of heredity that *The Verge* addresses). The subject also recurs in her biography of Jig Cook, *The Road to the Temple* (1927), the play *Alison's House* (1930), and Glaspell's last novel, *Judd Rankin's Daughter* (1945).

4. The eugenic message reached the American public through the combined forces of theater, film, and education (elementary through university), as well as fairs, expositions, displays, museum exhibits, conferences, lectures, and sermons, many of which overlapped as they circulated and shared materials and performance strategies. Eugenics also exploded onto the journal and newspaper scene after 1904, with dozens of periodicals carrying hundreds of articles; between 1900 and 1935, approximately five hundred books about eugenics were written for a general audience, and a 1924 bibliography of eugenics literature listed more than four thousand publications, about sixteen hundred of which were popular texts published in the United States between 1900 and 1924 (Holmes 1924).

5. On the tepid response to Mendel by his contemporaries, see Hartl and Orel (1992, 245–53).

6. Although biological determinism appeared to dominate eugenic public discourse, it was nonetheless colored by different degrees of neo-Lamarckian, or environmental views. I focus on hard-line eugenics because that is what Glaspell is contesting. For a discussion of the subtle variations between these views in eugenicists' arguments, see Hasian (1996), especially chapter 1.

7. The post-World War II struggle to designate eugenics a temporary aberration, a "pseudo-science" completely separate from the science of classical genetics, has been criticized by a number of scholars and deserves continued attention, especially given the present rhetoric surrounding the Human Genome Project. See Hasian (1996) for an analysis of retroactive constructions of eugenics, and Lewontin (2000) for potentially similar current trends towards what he calls "genomania," and the attempt to claim a value-free genetics removed from disturbing questions of human engineering.

8. Susan Glaspell, 1987, *The Verge*, in *Plays by Susan Glaspell*, edited by C. W. E. Bigsby, 57–101. All references are to this edition and will be cited in the text by act and page number.

9. The importance—and disturbingly unfamiliar aspect—of the plant is reminiscent of Strindberg, whom Glaspell had read and discussed with Cook. Moreover, flower imagery carries a similar weight in *Dream Play*. Strindberg also complemented his dramatic symbolism with his idiosyncratic experiments in natural history, which, taking his cue from Émile Zola, he recounted in his "Vivisections"—penetrating pieces aimed at emulating a dissecting scientist.

10. Glaspell felt deeply her own inability to have children. In her biography of her husband, she describes their mutual grief on learning that her health would prevent her ever having children (1927, 239).

11. The tower, which has a "back [that] is curved, then jagged lines break from that, and the front is a queer bulging window—in a curve that leans" (2. 78), evokes the use of space in the 1919 film *The Cabinet of Dr. Caligari*. Several critics have noted similarities between Claire's physical and emotional predicament and the situations facing the heroines of Charlotte Perkins Gilman's story, "The Yellow Wallpaper" (1892), and Kate Chopin's *The Awakening* (1899). See, for example, Kolodny (1985) and Payerle (1984).

12. As Nicole Hahn Rafter demonstrates, the altered names in the eugenic family studies illustrate this phenomenon (1988). Fifteen eugenic family studies, from article to book-length accounts of specific "defective" and "degenerate" families, were published in the United States from 1877 to 1926; two were so successful that they remained in print into the 1960s. In this popular literature, changed names routinely reduced humans to the level of animals or insects (e.g., Jake Rat, Muskrat Charlie, Bessy Spider, Woodchuck Sam), or simply indicated the core genetic rot from which all members were allegedly suffering (e.g., Rotten Jimmy, Maggie Rust). Surnames were also replaced by reductive nicknames (e.g., the Sixties family, named for the progenitor's supposed IQ) (Rafter 1988, 26–30).

13. The names of the characters also begin sequentially with the first letters of the alphabet: Adelaide, Breath of Life, Claire, (the dead David), Dick, Edge Vine, Edgeworthy, Elizabeth, Emmons, and Harry. In the alphabetical lineup, plants are again equal to human beings. Here, Glaspell uses letters to spell out ways in which the characters are distanced from, as well as linked to, each other. I am grateful to Robert A. Ferguson for this observation.

14. It is common knowledge that the hymn was also strongly associated with the 1912 sinking of the *Titanic*; therefore, use of this hymn would have suggested impending disaster and loss to Glaspell's audience.

15. These influences include Monist idealism, Nietzschean philosophy, aviation, feminism, Christianity, Strindberg's work, Jig Cook's passions, World War I, and psychoanalysis. A number of scholars have paid attention to one or more of the above, including Gainor, who examines how Glaspell's relationship with Cook affected her work (1989), and Noe (1995), Nelligan (1995), and Ben-Zvi (1989), who explore Glaspell's relationship to feminism. The only direct, although incomplete, account of the play's psychoanalytic suggestions belongs to Sievers (1955, 71).

References

Beeson, Emma Burbank. 1927. *The Early Life and Letters of Luther Burbank*. New York: Dodd, Mead.

Ben-Zvi, Linda. 1989. "Susan Glaspell's Contributions to Contemporary Women Playwrights." In *Feminine Focus: The New Women Playwrights*, edited by Enoch Brater. New York: Oxford University Press.

Bernard, Claude. [1865] 1949. *An Introduction to the Study of Experimental Medicine*. Reprint, translated by H. C. Greene. New York: Macmillan.

Bowler, Peter J. 1989. *The Mendelian Revolution: The Emergence of Hereditarian Concepts in Modern Science and Society*. London: Athlone Press.

Burbank, Luther. 1907. *The Training of the Human Plant*. New York: The Century Company.

Chesterton, G. K. [1922] 1987. *Eugenics and Other Evils*. In *The Collected Works of G. K. Chesterton*. Vol. 4. Reprint, with a forward by James V. Schall, S. J. San Francisco, Calif.: Ignatius Press.

"Claire—Superwoman or Plain Egomaniac? A No-Verdict Disputation." 1921. *Greenwich Villager*, 1 no. 21 (30 November): 1–4.

Darden, Lindley. 1991. *Theory Change in Science: Strategies from Mendelian Genetics*. Oxford: Oxford University Press.

Davenport, Charles. 1911. *Heredity in Relation to Eugenics*. New York: Henry Holt.

Fairs folder. 1920–1925. Cold Spring Harbor Series. American Philosophical Society, Philadelphia, Pennsylvania.

Gainor, J. Ellen. 1989. "A Stage of Her Own: Susan Glaspell's *The Verge* and Women's Dramaturgy." *The Journal of American Drama and Theatre* 1 (Spring): 79–99.

Glaspell, Susan. 1927. *The Road to the Temple*. New York: Frederick A. Stokes.

———. 1987. *The Verge*. In *Plays by Susan Glaspell*, edited by C. W. E. Bigsby. Cambridge, Mass.: Cambridge University Press.

Hale, Ruth. 1921. "Concerning *The Verge*." Letter to the Dramatic Editor, *New York Times*, 20 November, sec. 6, p. 1.

Hartl, Daniel L., and Vitezlav Orel. 1992. "What did Gregor Mendel think he discovered?" *Genetics* 13 no. 1 (June): 245–53.

Hasian, Marouf Arif, Jr. 1996. *The Rhetoric of Eugenics in Anglo-American Thought*. Athens: University of Georgia Press.

Holmes, S. J. 1924. *A Bibliography of Eugenics. University of California Publications in Zoology* 25 no. 2 (January): 1–514.

Kolodny, Annette. 1985. "A Map for Rereading: Or, Gender and the Interpretation of Literary Texts." In *The New Feminist Criticism: Essays on Women, Literature and Theory*, edited by Elaine Showalter. New York: Pantheon.

Lewisohn, Ludwig. 1932. *Expression in America*. New York: Harper and Brothers.

Lewontin, Richard. 2000. *It Ain't Necessarily So: The Dream of the Human Genome and Other Illusions*. New York: New York Review of Books.

Macgowan, Kenneth. 1921. "The New Play." *Evening Globe*, 15 November, n. p. Provincetown Players Collection, New York Public Library.

Nelligan, Liza Maeve. 1995. "'The Haunting Beauty from the Life We've Left': A Contextual Reading of *Trifles* and *The Verge*." In *Susan Glaspell: Essays on Her Theater and Fiction*, edited by Linda Ben-Zvi. Ann Arbor: University of Michigan Press.

Noe, Marcia. 1995. "*The Verge*: L'Ecriture Feminine at the Provincetown." In *Susan Glaspell: Essays on Her Theater and Fiction*, edited by Linda Ben-Zvi. Ann Arbor: University of Michigan Press.

Parker, Robert A. 1921. "Drama—Plays Domestic and Imported." *The Independent*, 17 December, 296.

Payerle, Margaret Jane. 1984. "A Little Like Outlaws": The Metaphorical Use of Restricted Space in the Works of Certain American Women Realistic Writers. Ph.D. diss. Case Western Reserve University.

The Provincetown Plans at Their Experimental Theater. 1921. *New York Call*, 30 October, n. p. Provincetown Players Collection, New York Public Library.

Rafter, Nicole Hahn, ed. and introduction. 1988. *White Trash: The Eugenic Family Studies 1877–1919*. Boston, Mass.: Northeastern University Press.

Rathburn, Stephen. 1921. "*The Verge*, the Provincetown Player's First Bill, Is an Extraordinary Study of the Superwoman." *The Sun*, 19 November, 4.

Sievers, Wieder David. 1955. *Freud on Broadway*. New York: Hermitage House.

Wiggam, Albert Edward. 1924. *The Fruit of the Family Tree*. Indianapolis, Ind.: Bobbs Merrill Company.

Young, Stark. 1921. "After the Play." *The New Republic*, 7 December, 47.

Helen in Philadelphia: H. D.'s Eugenic Paganism

Andrew Lawson

EZRA POUND MARKED THE OCCASION OF HIS RETURN TO AMERICA IN 1913 by writing an essay entitled "Patria Mia." In the essay, Pound declares that it is a mistake to think that the United States "is by race Anglo-Saxon." The American "race," Pound insists, is "mongrel"—"one stock neutralizing the forces of the other" (1962, 12). The "surging crowd" on Seventh Avenue, New York, is "as pagan as ever imperial Rome was." To Pound's eyes, these "mongrel" Americans move with "an animal vigour;" their gods "are not the gods whom one was reared to reverence" (14). Worse still, "one knows that they are the dominant people and that they are against all delicate things" (14). Pound's account of racial-cultural degeneration is familiar enough, following what, after Henry Adams's *Education* (1907) and Henry James's *The American Scene* (1906), is a veritable tradition of exiles who return to find that history has passed them by and the world turned unutterably strange. But Pound goes on to make a striking prediction: "[a]fter the attempted revival of mysticism," he says, "we may be in for a new donation, a sort of eugenic paganism" (13).

Pound's sense of racial degeneration and his anxious search for cultural moorings are recapitulated in H. D.'s autobiographical novel of 1927, *Her* [*HERmione*]. The novel provides a fictionalized account of Hilda Doolittle's youth in Philadelphia, covering the period from 1906 to 1911, in which she dropped out of Bryn Mawr, met and became engaged to Ezra Pound, fell in love with Frances Gregg, suffered a nervous breakdown, and left, with Gregg, for Europe. A modernist bildungsroman, *Her* describes the emergence of a would-be "autonomous" self from a tangled matrix of social conventions. *Her*'s quintessentially modernist project is to make the self a source of power rather than a site of determinations. But the novel is also immersed in historically specific discourses of class and race, which it both obscures and reveals, namely, the discourses of American eugenics and Paterian Hellenism. In reading *Her*, I want to describe both the vertiginous spaces that these discourses open up for negotiation by the modern self, and the ideological limits inscribed in them. In particular, my concern is with what the drive to autonomy conceals, with what Judith Butler calls the "conceit" of autonomy, its "unmarked class character."[1]

The protagonist of H. D.'s novel is Hermione Gart, whose name for her authentic, 'inner' self is Her. Hermione has returned to her parents' home in a rural suburb of Philadelphia, "drowning" in a state of "dementia," after failing in mathematics at the elite women's college, Bryn Mawr (1984, 2). The novel recounts a search for markers that might fix Hermione's identity: "[s]he must have an image no matter how fluid, how inchoate" (5). This search is made problematic by a lack of cultural foundations: Pennsylvania is "primaeval," the classic American wilderness. Accordingly, Hermione clings "to small trivial vestiges, not knowing why she so clung" (8). Pennsylvania holds her "to things of no value, small totems that meant some tribal affinities with European races" (9). Hermione's modernist desire for an "image" is related to what John Higham has described as American nativism: the quest of native-born, upper-middle-class Americans for origins in an effort to maintain cultural authority in the face of polyglot immigrant hordes on the one hand, and vulgar, Gilded Age plutocrats on the other.[2] The opening section of *Her* is quite explicit, if a touch ironic, about Hermione's nativist class consciousness. Hermione thinks of her neighbors, the Farrands, as "people with too much money," while her mother, Eugenia, "tolerantly" describes them as "really nice," despite being "business people" (12). Apart from the parvenu Farrands, Hermione is said to know "the very best people," who, in Philadelphia, are "distinguished" by their "European affinities" (12). "Affinity" is a highly charged word in Hermione's vocabulary, carrying the sense of "related by marriage," as well as the more specialized sense of "spiritual attraction between persons" (*OED*). In its discovery of the latter sense, specifically in Hermione's spiritual-sexual attraction to Fayne Rabb, the novel looks for "other kinships" (57) than those of the "Anglo-saccharine backwash of Bryn Mawr" (129) with its "insular and specialized race consciousness" (57).

"Anglo-saccharine" is Hermione's ironic play on "Anglo-Saxonism," which Higham describes as the "complacent, self-congratulatory" discourse of the New England elite who saw themselves as the guardians of the wellsprings of American civilization (1963, 134).[3] They constituted the sturdy Germanic stock eulogized by Matthew Arnold as "an incomparable, all-transforming remnant," like the remnant of Israel prophesied by Isaiah (1885, 70). By the 1900s, there were increasing fears that this saving remnant might be threatened with extinction, that the higher fertility of immigrant populations would result in "race suicide" (142–47). Progressivist reformers at the University of Pennsylvania, led by Susan P. Wharton, were, in David Levering Lewis's words, "prey to eugenic nightmares about 'native stock' and the better classes being swamped by fecund, dysgenic aliens" (1993, 188). For white, middle-class Philadelphians, the barbarians were at the gates.

In *Her*, "Anglo-Saxonism" as the stuff of genteel pretension and casual snobbery is a thoroughly alienating and antiquated discourse. But this doesn't mean that Hermione effortlessly transcends the "specialized race-consciousness" of Old Philadelphia in a radical gesture of modernist revolt against social determinations (H. D., 57). *Her* attempts to free its protagonist from the banalities of class and race consciousness, while holding on to a sense of distinction.[4] In doing so, it repeats in an even more virulent form the patterns of social hierarchy and exclusion that Anglo-Saxonism, Philadelphia-style, is devoted to maintaining.

A Creature of Ebony

The person Hermione is obliged to call "sister" is not her sister, but her sister-in-law, Minnie Hurloe, whose very presence "depreciate[s] the house" (15). While the Garts are "Nordic," or "New English," with a "Germanic affiliation," the "nonchalant and aggressive" Minnie is only "some two generations" rooted in America (46). With her lack of ancestry, Minnie is more typically "American" than the Garts: she is a product of "another centre," associated with the Midwest, modernity, and jazz (46). Paradoxically, it is the Garts who are the "alien" growth in modern America, literally an endangered species. *Her* is the story of a search for another, more authentic sister, "a creature of ebony" or "an image of ivory" who will transcend both the banalities of Old Philadelphia and the vulgarities of the parvenue Minnie (10). But Hermione's fantasy of sisterhood is from the outset bound up with a racialized class consciousness in which spiritual affinities settle into hierarchical distinctions.

The "creature of ebony" is the family's black maidservant, Mandy, referred to approvingly by the Garts' neighbor, Mrs. Rennenstocker, as a "darkie" of "the real old Southern kind" (115). In the kitchen with Mandy, Hermione "slip[s] a white hand into the deep bowl," and a "black arm lift[s] from the deep bowl" (27). The image posits a kind of fusion of black and white selves, a melting pot in which the one is the inverted image of the other, held for a moment in the tight reciprocity of the gaze. Hermione identifies with Mandy, rather than with Minnie, who says "I know these darkies are so dreadful" (39). "Mandy is adorable," she thinks, "I adore Mandy. How Mandy ha-aates Minnie" (39). Hermione adopts the discursive practice described by George Frederickson as Negrophilia, a practice central to the pleasures and anxieties of "Anglo-Saxon" self-fashioning. Here, the "white desire" for embodiment is passed through "a fictive veil of blackness," allowing Hermione to achieve both a corporeal presence and a mark of distinction, a spiritual affinity beyond the vulgar Minnie Hurloe (Neilsen 1988, 28).[5] But Hermione's identification with Mandy has its structural limits: the practice of Negrophilia is paralleled by an equally intense Negrophobia.

Returning home late from an attempt at seduction by the Ezra Pound figure, George Lowndes, Hermione is challenged by Eugenia to explain her erratic behavior. Eugenia tells Hermione she is hurt by her calling her by her first name. Hermione begins to justify herself by explaining, somewhat tendentiously, that it would be confusing if everyone in the family were referred to by their surname, Gart. In that case, she says, Minnie's dog Jock would be "rightful Gart and Mandy . . ." (78). Having identified with Mandy to the extent of a metaphorical fusion of black and white hands in a mixing bowl, Hermione sharply distances herself, to the extent that Mandy is equated with Minnie's dog in her lack of entitlement to the family name. The would-be transgressive fusion of self and other across the borderline of race is succeeded by a reiterated claim to ownership: "Mandy belongs to us. Mandy belongs to me. Mandy is mine" (78). This drive to appropriate the other is revealed to have a political motivation in a speech by Hermione which, taken out of its historical context, is remarkably obscure: "This business of the United States, United States of America doing away with states being separate with separate states and each state with its own laws is what is responsible for all this mob rule. . . . You get no sort of cohesion out of a thing so immense. . . . This thing that anyone can say *united we stand* is all rot. We can't stand united. Divided we would probably stand" (78). What connects Hermione's claim to own Mandy with her invective against federal law is, I think, a contemporary political claim by the other speaking with its own voice.

On 14 August 1908, in Lincoln's birthplace of Springfield, Illinois, a white woman, Mrs. Earl Hallam, accused a black man, George Richardson, of attempting to rape her. There followed two days of rioting and lynching by a white mob, in what the reporter William English Walling called "The Race War in the North." English's article in the liberal newspaper the *Independent* called for a "large and powerful body" of Northern whites to "come to [the] aid" of the Negro (534). In May of the following year, at the National Negro Conference organized by a group of Progressives and socialists in New York, W. E. B. Du Bois called for the restoration of legal and political equality for African Americans. The Conference resolutions appealed for the Constitution to be enforced and civil rights to be guaranteed under the Fourteenth Amendment, "equal educational opportunties" for all Americans, and the "right of the Negro to the ballot" guaranteed by the Fifteenth Amendment (Kellogg 1967, 302–3). Hermione perceives the civil rights claimed by Du Bois and the Niagara Movement under federal jurisdiction as a threat to national unity: the unity of the nation is best assured by leaving states to run their own business. Holding the federal assertion of rights to be responsible for mob rule, Hermione follows the political logic of segregation, asserting her right to own her servant, as if Philadelphia were a Southern plantation.[6]

This historical conjuncture of a militant claim for full civil rights and the reassertion of white paternalism significantly complicates critical claims made for *Her*'s depiction of Mandy as a recognition of the other, a straightforward identification with a global principle of "difference."[7] What is involved in Hermione's identification with Mandy is the ambivalence described by Homi Bhabha as both a "recognition" of difference and a "disavowal" of it (1994a, 75). Here, "difference" is another name for the self-image looked for, and perceived in, the other. This looking-for-difference constitutes "the narcissistic demand" that the other confer authority on the self (Bhabha 1994b, 97). Hermione's narcissistic identification with Mandy deflects any resemblance she might have to the deracinated and vulgar Minnie Hurloe. Negrophilia functions, in a highly contingent manner, to reinforce the "Nordic" distinction of Old Philadelphia. But there exists a residuum of hostility to the other in the very act of identification: hostility produced by the inconvenient ways in which the mirror image fails to reflect the self adequately. Hermione's Negrophilia requires and assumes an "unchanging" other, who is "absolutely different," ruling out historical change and self-determination (Said 1995, 96).[8] But Mandy has become a risky investment: the symbolic currency of a long historical association between white mistress and black maidservant has been devalued by increasingly militant claims for rights. The contingency of these claims puts enormous stresses on Hermione's psychic identifications. Indeed, in developing her political theme, Hermione goes further: "You're defying laws of science . . . mathematics and chemistry by trying to mix such mobs heterogeneously. You can't expect things to go on forever this way. You'll get mob rule and then mob rule and then mob rule" (78–79). Hermione calls on the "laws of science" to resist the dangers of racial mixing and cultural heterogeneity, signaling the adoption of another discourse with a lasting influence, American eugenics.

A Breeding Experiment

In 1908, while the race war was raging in Springfield, Illinois, William Ripley offered the readers of the *Atlantic Monthly* a theory that answered an urgent contemporary danger: the danger that the crossing of "superior" and "inferior" races produced, potentially, by mass immigration, would drag the "superior" races down. Drawing on the discoveries of Mendelian genetics, Ripley warned that there was a real threat of "reversion" to the original stock if "the crossing of types" was "too violently extreme" (751–55). In other words, blond Teutons would revert to swarthy Mediterraneans if native Irish stock were to breed with immigrant Italians. This eventually became the message to the nation of the "eugenics cult" that swept America in the 1920s.[9] Charles B. Davenport lamented that "a hyberdized [sic] people are a badly put together people and a dissatisfied, restless, ineffective people."[10] Madison Grant, in *The Passing of the Great Race* in 1916, warned of the "many amazing racial hybrids and some ethnic horrors that will be beyond the powers of future anthropologists to unravel" (1918, 92). In this state of emergency, according to Davenport, "race homogeneity" and immigration restriction were the only "safe course" (Pickens 1968, 57). What Hermione is claiming, with her appeal to the "laws of science," is the right to own Mandy as a bulwark against the spread of amazing hybrids. Owning Mandy, unambiguously black in the stark visibility of the stereotype, will keep the lines on the racial map clearly drawn, combating both the agitation for civil rights, and the heterogeneity threatened by mass immigration.[11]

The discourse of eugenics makes its appearance in the opening section of *Her* precisely when the racialized class consciousness of Anglo-Saxonism is being treated most sceptically. As she makes her way through the woods to her ancestral home, Hermione laments the fact that Pennsylvania's rivers hold "nothing" in the way of historical associations or myths: "she wanted sand under bare heels, a dog, her own, some sort of Nordic wolfhound" (H. D. 1984, 6). The narrator points out that Hermione could not have known "that Pennsylvania bears traces of a superimposed county-England and of a luscious beauty-loving Saxony. She could not know that the birdfoot violets she so especially cherished had far Alpine kinsfolk" (9). The terms "Teuton" and "Alpine" had been coined by Ripley in *The Races of Europe* (1899) as part of a pioneering effort to distinguish the "old" Northern European immigration from the "new" influx from the Southern and Eastern countries of Europe. But *Her*'s narrator applies the later findings of Grant, who separates the "races" of Europe and assigns them in a hierarchy: the "Alpines" of central Europe are "submissive" peasants (1918, 227), Mediterraneans are physically weak and artistic, while the Nordics are a race of "rulers, organizers, and aristocrats . . . domineering, individualistic, self-reliant and jealous of their personal freedom" (228)—"the white man par excellence" (167). The population of America at the time of the Revolutionary War was "overwhelmingly Nordic" (83) but is now threatened by Alpines, Mediterraneans, and Jewish hybrids: what Grant calls a "native American aristocracy resting upon layer after layer of immigrants of lower races" (5). Unabashedly antidemocratic, Grant makes class consciousness dependent on racial purity of a particularly refined kind: a version of white supremacy tailored to the "specialized race consciousness" (H. D., 57) of the New England remnant.[12] Hermione summons the cutting-edge race-thinking of Ripley and Grant to establish a still deeper "affinity" than Anglo-Saxonism, a "Nordic" affinity that is both spiritual and biological.

Hermione's identification with eugenic discourse is, however, split and contradictory because it is, in turn, identified with her father, Carl Gart. Hermione decides her father is "Athenian," while her mother, Eugenia, is "Eleusinian" (31). Eugenia is associated with the Eleusinian mysteries, with the cult of Aphrodite and Dionysus, and with the wildness of the passions. Carl Gart is

Athenian or Apollonian in his devotion to the ideal, to logic and order—absorbed in his studies, he is incapable of communicating with his daughter. During a thunderstorm that produces an eerie, green light in the room, Eugenia and Hermione, mother and daughter, are thrown into "profound intimacy" and reconciled after the mythical pattern of the mother Demeter finding her daughter Persephone in the Underworld (88). In this scene of reconciliation, the rain floods the father's aquarium, destroying his five-year "breeding experiment" (92). The eugenic anxiety of the Nordic patriarch about "breeding" is, seemingly, overcome by the pre-Oedipal union of mother and daughter: eugenics is displaced by Eu-genia, fulfilling Hermione's vow that "mythopoeic mind (mine) will disprove science and biological-mathematic definition" (76).

But this identification with the mythical mother against the scientist father is itself problematic, since Eugenia is the parent most obviously concerned with "breeding" in the sense of maintaining kinship boundaries and hierarchies: for her, the Farrands are "business people," and the bohemian George Lowndes will not do as a marriage-partner for her daughter. Hermione rejects eugenics as "merciful surveillance," in Sir Francis Galton's terms: a surveillance aimed at ensuring the marriage of "like with like" and an enhanced stock of "cultured families" (Galton 1909, 20). But Hermione embraces the language of eugenics as a means of modernist self-fashioning. Nordic hounds and Alpine violets can be summoned as images of mastery and pleasure, cognate with the deeper, "autonomous" affinities of the self, even as the hopes of the father's merely imitative and routine "breeding experiment" are dashed. Crucially, this tactic of adapting a discourse in order to authorize the self proceeds without disturbing the discourse's ideological valence. Hermione's "Nordic" affinities are developed in a privatized space of "consciousness" that is contiguous with, and parallel to, the political project of eugenics. This contiguity is preserved by Hermione's adherence to her class location within the embattled New England remnant, even as she refashions its cultural forms of self-expression. The same tactic of a privatized rebellion within an established framework can be seen at work in the development of Hermione's concept of "mythopoeic mind."

An Image of Ivory

Hermione recruits Mandy in her battle against the banalities of Old Philadelphia. Sitting at breakfast with her mother in the charged atmosphere produced by the thunderstorm, Hermione exclaims "[w]e're all green like faces under water"—observing that only Mandy is "different": "Mandy (exquisite bronze) was a brazier burning in that bleak room. . . . Eugenia was shushing at Hermione, not wanting her to say it. I can't say Mandy is a bronze. I can't say Mandy looks like Etruscan bronze dredged from the mid-Ionian with colour flashing against her polished bronze. . . . I must say, 'What Mandy—not more hot cakes?'" (88). This image of Mandy as an archaic statue appears to confer an aesthetic privilege on racial difference, by which Mandy's skin color, "exquisite bronze," is set off in ironic contrast to the drab gentility of the surroundings in which she has a menial function. Mandy appears to have been translated from the colored mammy stereotype into a new space of shared distinction, her aesthetic qualities equivalent to the gaze that discerns them. But Hermione's vocabularly is derived from a discourse that steers her perceptions in a direction that makes difference the basis of racial hierarchy, rather than equivalence. Hermione's image of Mandy is "dredged" or recovered from Walter Pater's *Greek Studies* (1925a), a text that is part of the larger, cultural project of late Victorian Hellenism.

Pater traces the action in Greek history of what he calls the "Dorian" and the "Ionian" tendencies. Initially, this looks like an innocent opposition between "Dorian" intellect and "Ionian" sensuality, an ethical "abstractness and calm," set against an aesthetic delight in "brightness and colour, in beautiful material, in changeful form everywhere" (Pater 1925a, 253, 252). But Pater's terms are not merely neutral lenses for the specular consumption of "otherness"—they have a historically charged, specifically racialized connotation. They derive from Karl Otfried Müller's *History of Greek Tribes and Cities* (1820–24), which set out to prove that Greek civilization had not been influenced by the "Ionian" cultures of the Orient, as some of the Ancients themselves thought, but was purely Western or "Dorian."[13] Thus, Pater insists that the "Dorian, European, Apolline influence" is "the true Hellenic influence" (256), and he finds this influence triumphantly celebrated in the Parian marbles discovered at the temple of Aphaia on the island of Aegina in 1811 (Pater 1925a). The marbles depict a "combat between Greeks and Asiatics concerning the body of a Greek hero, fallen among the foemen" in the Trojan War (258). For Pater, the marbles function as "an embodiment of the Hellenic spirit," which is set against the "gaudy spectacle of Asiatic, or archaic art" (263). What is striking in Pater's account is its use of the present tense, giving a sense of imminent racial threat, an always impending showdown in the continuing struggle between Western "liberty" and Eastern "barbarism." The Aegina marbles provide a "revelation . . . of the temper which made the victories of Marathon and Salamis possible" (260). But they also herald the "macabre" approach "of those lower forms of life which await to-morrow the fair bodies of the heroes" (266).

Hermione's description of Mandy as "Ionian," then, is a crucial refinement of the "coloured mammy" stereotype of romantic racism or Negrophilia. It is not simply that "difference" is produced in a scene of recognition of the other. Instead, a racial hierarchy is established in terms whose valence within the discourse of Hellenism has been long established. Mandy is positioned at a point of cultural intersection, where "Western" rationality meets "Eastern" sensuality, a sensuality at one remove from barbarism. Mandy's otherness serves to differentiate Hermione from the banalities of Old Philadelphia. But this gesture of recognition is also one of disavowal: Mandy is objectified as a cultural object positioned outside the proper boundaries of culture.[14] As the stresses imposed on the would-be autonomous self increase, with Hermione steered by her parents towards what must be a suitable marriage, the hierarchy of "difference" is asserted in an increasingly extreme and violent struggle for selfhood. From this point in the narrative, the discourses of racism, eugenics, and Hellenism overlap and interfere, reinforcing each other's exclusions and distinctions.

Hermione receives two complimentary tickets from Fayne Rabb to a production of *Pygmalion* in which she, Fayne, is to take the leading role. At the end of the corridor where she and George are sitting awaiting the performance, Hermione sees an image of "a naked woman standing on a sea shell." This "dumpy," "too white" Aphrodite resembles the "cheap mother-of-pearl handles to little showy cheap knives, like the mother-of-pearl top to a workbox Mandy had, that Mandy loved" (H. D. 1984, 136). Deciding "that sort of thing's all right for Mandy," Hermione exclaims: "that sort of woman standing on a sea shell sort of art is simply negroid" (136). Mandy's affinity with the "gaudy spectacle" of Asiatic art is thus firmly established. Hellenism has been cheapened and contaminated in its commodified, mass cultural form. It is necessary for this kitsch Hellenism, which can become anyone's property, even the property of Negroes, to be distinguished from that "mellow humane super-European consciousness . . . something greater that went with planets swirling" (137). In *Her*, the privilege of conferring identity in difference (Mandy as "Her's," Mandy as Etruscan bronze), has been disturbed by the racial other's identification with "white" culture.[15] Illicit affinities are claimed in the modern marketplace, a space of promiscuous commodification and exchange, where racial and class distinctions are not necessarily respected. This danger is overcome when "a girl in a Greek tunic" stands "for one ironic moment before the negroid sort of art-picture of a woman on a sea shell" (138). Fayne literally effaces Mandy's "negroid" image, becoming the authentic Aphrodite, the sister as "image of ivory," stepping into the space of the other once occupied by Hermione's servant.

Hermione has learned that "[a]rt was the discriminating and selecting and bringing odd distorted images into right perspective" (139). She knows that, in a switch to a fiduciary metaphor, "there are gamblers of the spirit as there are gamblers of the mind"—that spiritual capital is available "to hoard or to risk in one reckless spendthrift moment" (52). The identification with Mandy's "otherness" has, in effect, been foreclosed, in Judith Butler's terms: a symbolic investment has been weighed in the balance and found wanting (Butler 1993, 190). If Minnie were an unsuitable sister, a "common creature" (H. D., 195) of only two generations' poor-breeding, Mandy has been ruled out by her lack of cultural distinction and her race's troubling assertion of equal rights. Hermione has to foreclose on Mandy in order to preserve cultural distinction and authority: in this sense, there is every reason to be "discriminating." Reassuringly, Fayne is not only Aphrodite, the ivory image of desire, she is Nordic to boot: Eugenia informs Hermione when she receives the tickets to the play ("Compliments of Fayne Rabb") that the Rabbs were "originally Dutch" (128).

With the full assumption of this cultural authority, the Negrophilia of Hermione's initial identification with and ownership of Mandy is replaced by the rival logic of Negrophobia. Hermione is "sustained" by George's Hellenistic description of her verses as "[c]horiambics of a forgotten Melic" on the streetcar to Fayne's West Philadelphia home, when she finds herself pressed "against the bulk of a huge negress," who boards the bus "lumbering," with a "harsh bull-bellow" (149). The "intelligible" body of Mandy, the "creature of ebony," has given way to the "abject, unliveable" body that, again in Butler's words, "haunts the former domain as the spectre of its own impossibility, the very limit of its intelligibility, its constitutive outside" (1993, xi). Mandy's "creaturely" body, with its connotations of degrading labor and cultural abjection, confirms the distinction of the aloof image of ivory, in a "fantasy of mastery" (Butler 1993, 190, 244 n. 7). It is this fantasy of mastery that *Her* proceeds to put to the test, charting the spiritual-sexual affinities between Hermione and Fayne, sibling members of an endangered elite.

Tell the Lacedaemonians

In her relationship with Fayne, Hermione practices a thoroughly Paterian ascesis or self-mastery, with desire controlled and sublimated in the intellectual enjoyment of the body as an aesthetic object of contemplation.[16] Hermione tells Fayne that she "might have been a huntress," and pictures her body "[b]eneath schoolgirl blue serge" as a Greek marble from Heliopolis (H. D. 1984, 163, 162). The climax of the relationship comes when Hermione places her hands over Fayne's eyes and Fayne tells her: "Your hands are healing. They have dynamic white power" (180). Hermione feels "[f]ire and electric white sparks puls[e]" in her "thin wrists," and Fayne falls asleep, creating a minor scandal in the Gart household (180). The scandal is unnecessary, since there are, Hermione knows, "passions of the psyche as well as passions of the body" (52). "Greek sensuousness," in Pater's words, "does not fever the consciousness" (1986, 142).[17]

If it is impossible to disentangle this "white eroticism" from its racial connotations, from the Dorian "dream of white friendship," then it must also be acknowledged that this eroticism has a no less complex, and contradictory, eugenic significance (Gregory 1997, 98). Mastery of the passions is required by those responsible for breeding with the best: "merciful surveillance" is also self-surveillance. Another reason to foreclose on Mandy is that Negro-Asiatic sensuality, or what Pater calls "languid Ionian voluptuousness," threatens the secure self-possession entailed by membership of a cultured elite (quoted in Gregory, 88). What H. D. elsewhere refers to as "Asiatic abandon," signaling the delusive snares of a throughly racialized sensuality, is abjured in favor of a "Doric" chastity and stoicism.[18] The physical boundaries of the self are also psychic borders, and they must be subjected to "merciful surveillance." Whiteness is bred, as it were, in the bone: it comes to stand for the inviolate inner sanctum of the self, a self-possession that is linked to the possession of a homeland. This homeland has to be defended against both the figural incursions of desire, and the barbarian invaders poised more literally on its borders. New identities are needed for this defense.

Hermione is transformed by the white eroticism of Paterian Hellenism, but into something other than mere flesh and blood. When George Lowndes makes another of his clumsy attempts to seduce her, this time on her parents' sofa, Hermione escapes "[s]traight and strong like some girl athlete from Laconian hill slopes, straight and brave like the maiden Artemis" (H. D. 1984, 175). Talking with George about their planned marriage and removal to Europe, Hermione becomes "an external objectified self, a thin vibrant and intensely sincere sort of unsexed warrior" (187). This ambiguous image of a girl athlete who is also an unsexed warrior forms part of a chain of images that begins with the sister, the image of ivory, "lost on some Lacedaemonian hill" (10). The image combines two alternative feminine identities: the armed Athene, representing intelligence and will, and Artemis, who stands for a chaste, yet erotically charged wildness. But what is crucial to the eugenic discourse valorized by *Her* is that this image is linked with Lacedaemonia, otherwise known as the Greek city-state of Sparta, through another appropriation of Pater's Hellenism.

For Pater, Sparta is "the Dorian metropolis" (1909, 198), the citadel of Greekness, organically connected with "the old roots of national life" (202). Undertaking an imaginary journey to the city with Müller as tourguide, Pater provides tantalizing glimpses of naked Spartan youths exercising on the banks of the Eurota, in "wholesome vigour . . . clearness and purity" (214). In Sparta, Pater affirms, youth enjoys "that kind of well-being" that comes from "the order and regularity of system," from "living under central military command" (204). In this city of ultimate ascesis, even Aphrodite is armed.[19] The beauty of Sparta, Pater asserts, "was a male beauty, far remote from feminine tenderness," and the exclusively masculine borders of the Spartan ascesis are rigorously patrolled by Pater's discourse (Rawson 1969, 363).[20] *Her*'s production of the androgynous "girl athlete" or "unsexed warrior" is, in this context, a stunning appropriation of the martial discourse of Victorian *paiderastia*.[21] But this appropriation is made in order to transcend the body and sublimate desire: to preserve the self in a form that is specifically racial and endangered. For the point about Sparta is that it is a fallen city, overcome, eventually, by a foreign enemy. The fall of Sparta is a warning from across the centuries that carries a eugenic message.

Hermione tells Fayne that she doubts whether, as Americans, they are really "European," or merely "deracinated Europeans holding valiant intellectual standards" (218). She goes on, "We are held here to the thin slope of the Atlantic seaboard like the warriors of Leonidas. We hold, holding to our intellectual standards a sort of mountain. . . . A sort of weltgeist that has a vibrant phosphorescent heart, the so few of us, the so very few of us *Americans*. . . ." (218). Hermione refers to the famous story, told by Herodotus, of how three hundred Spartans led by their King, Leonidas, chose to die in a glorious last stand against the invading Persian army at Thermopylae in 480 B.C.: a noble defeat that bought the Greeks time to secure a final victory at Plataea the following year. Thermopylae and Troy alike signify a fundamental racial and cultural divide. It was for the sake of a "single Lacedaemonian girl," Helen of Troy, that "the Greek race" joined together in the invasion of Asia, with its "tribes of barbarians" (Herodotus 1910, 3). Thermopylae is another moment in a continuing struggle that symbolizes the superiority and cohesion of the West against the barbarian East. For another Victorian Hellenist, John Addington Symonds, Thermopylae's "lesson of a mortal peril escaped miraculously . . . quickened the spiritual conviction of the race" (1902, 381).

It is as the bearer of this quickened spiritual conviction that Hermione imagines herself in *Her*'s closing pages, in the guise of a Spartan girl athlete. There are good historical reasons for fixing on this image. Sparta was famous for the relative autonomy and status of its women citizens by contrast with Athens: girls were given gymnastic training alongside boys, while Spartan wives could own property and were renowned for their outspokenness, loose attire, and physical prowess.[22] But the "physical training for the female," admired by Xenophon, was prescribed in the interests of producing "more vigorous offspring." The role of Spartan women was adapted to preserving what Plutarch calls Sparta's "noble seed," its closed aristocracy and standing army (quoted in Cartledge 1981, 91). As Paul Cartledge notes, the relative autonomy of Spartan women was designed to serve "exclusively eugenic ends" (92). This practice did not, however, guarantee Spartan hegemony. Because of its dependence on small hereditary holdings of land and laws excluding outsiders, Sparta's citizen population suffered a catastrophic decline in the fourth century, and it was unable to resist the ascendancy of Thebes, confirmed at Leuctra in 371, where four hundred of the remaining seven hundred Spartans perished. The sacrifice of the noble few at Thermopylae is both Sparta's finest hour, and its eugenic nightmare, a premonition of its eventual decline and fall.[23]

Hermione's adoption of the image of the Spartan girl athlete in *Her* appeals directly to this eugenic context. The final, ideological function of Hellenism in *Her* is to reassert the primacy of the endangered Nordic race, the New England remnant—reimagined as the fallen heroes of Sparta. This is the message Hermione tells herself she is destined to carry "in forgotten metres," like Pheidippides, the Athenian who ran to alert Sparta to the Persian landing at Marathon (H. D. 1984, 220). But the message, rather than being a purely Hellenic one, is that "we are a creature even now seething with life cells, phosphorescent cells; will Fayne Rabb desert me. Run, run little blood corpuscle, tell the whole inchoate mass (Dakota, Oklahoma) that we are all together. . . . Tell the Lacadaemonians . . . that . . . we . . . lie . . . here . . ." (220–21). Hermione's message ends with Simonides's epitaph on the Spartans at Thermopylae: "O passer by, tell the Lacedaemonians that we lie here obeying their orders."[24] But the runner at Marathon has become a corpuscle in the endangered blood of the Nordic elite, the "biological mathematic definition" of the Athenian father merging with the "mythopoeic mind" of the Eleusian mother. *Her* delivers on Pound's prediction. Its message from the fallen Spartans to the inchoate, American mass is a modernist hybrid: eugenic paganism—the last gasp of a specialized class consciousness, an exhausted "race."[25]

H. D.'s novel begins with a meditation on home, on how to be at home in America, land of multiple origins and effaced peoples. Love of the nation as home secures the boundaries of the self in a series of affective ties: genealogies, affinities, and affiliations. For there to be a homeland there must be a stranger, an outsider, a figure of radical otherness who will both stabilize the boundaries of national identity and summon the productive anxiety that attends threshold states of being, since boundaries can be crossed, identities lost, definitions blurred. The outsider must be welcomed into the home just far enough to make it *feel* like home—and summarily banished when he or she makes as if to take up residence. This was Du Bois's insight, turned to stunning rhetorical effect in *The Souls of Black Folk*, when the "swarthy spectre" that sits "at the Nation's feast" eventually rises to assert, in its own ambivalent tones of defiance, anger, and overwhelming love, that it has been at home longer, and more passionately, with its "gift of the Spirit," than the owners of the house (1989, 5, 187).

It is, finally, politically dangerous and historically inaccurate to assert "ambivalence" as the psychological master trope of H. D.'s fiction: as though *Her* presented Mandy as the stereotyped equivalent of both adored figure of otherness and domestic servant, Ionian bronze and family pet, without making any final choice or investment in these images. The ambivalence towards the outsider in *Her* is inextricably bound up in a discourse with an ideological history, a discourse of "race" that is, at the same time, an assertion of class privilege, a means of self-fashioning that entails elite group membership. In *Her*, ambivalence and indeterminacy serve only to facilitate the pleasurable game of negotiated identity played by gamblers of the spirit. Mandy is banished from the house *in fantasy*, the occupants lie vanquished *in fantasy*. That is the modernist writer's privileged form of abjection, to fabricate and meditate on fictions. But the American nation's boundaries, although agitated, remain intact: and so do the social relations—degrading labor, cultural exclusion—that sustain them.

Notes

1. Butler 1993, 228; 227. Butler does not pursue this point, however.

2. Higham 1963. Walter Benn Michaels coins the term "nativist modernism," which he sees as complicit in the project of "racializing the American" (1995, 1, 13). While I agree with Michaels that the discourse of race works to "replace class difference with racial difference," my emphasis is on the disguised class interests that manifest themselves within the appeal to "race" (57).

3. See also Solomon 1956, chap. 4, and Klein 1981, chap. 1.

4. For Bourdieu (1986), distinction is the symbolic "profit" yielded by the "exclusive appropriation" of culture through unequally distributed "dispositions and competences." This enables culture to function as "cultural capital, i.e., as an instrument of domination" (228).

5. On Negrophilia and Negrophobia, see Frederickson 1987, chaps. 9, 10. On "whiteness" as incorporeal spirit versus embodied "blackness," see Dyer 1997, chap. 1.

6. On the "Negro caste practices" of nineteenth-century Philadelphia, see Warner 1968, 126. Du Bois, in *The Philadelphia Negro* of 1899, describes domestic service, as a "relic of slavery" (1967, 137). Isabel Eaton's "Special Report on Negro Domestic Service," which appeared as an appendix to Du Bois's study, found that ninety-one percent of "colored workingwomen" in Pennsylvania

were "in service" (428), the majority of them "supplied" by Southern states (436).

7. Du Plessis argues that "[r]acial difference, maternal mirroring and lesbian sororal passion are related—and positive—forms of Otherness in this novel" (1986, 68). Friedman reads the passage as a "lyric celebration of the bond between white and black," which allows Hermione to "defin[e] her difference from her conventionally feminine sister-in-law" (1987, 219; 220). The structure of identification Du Plessis and Friedman valorize here is problematic, however, since it involves a blindness to the function of the racial stereotype, and to its historical context.

8. See also Abdel Malak on the "hegemonism of possessing minorities" characteristic of Orientalism (quoted in Said 1995, 97).

9. On the American eugenics movement, see Haller 1963 and Ludmerer 1972; on the links between the American and British eugenics movements, see Kevles 1985, chaps. 4–7. See also Darrow 1926, 129–137.

10. Davenport, quoted in Pickens 1968, 57.

11. On the "status anxiety" motivating the need for the stereotype's "hard lines and clear concepts," see Gilman 1985, 20, 19. Gilman is quoting from an epigraph in Mary Douglas 1966, *Purity and Danger: An Analysis of Concepts of Pollution and Taboo.*

12. See also Higham 1963, 155–57.

13. On Müller, see Bernal 1987, 308–316.

14. On this Arnoldian sense of "culture" as a "system of exclusions" that designates a boundary, see Said 1983, 8–12.

15. Deriding the "widespread and fatuous belief in the power of environment, as well as education and opportunity to alter heredity," Grant states, in *The Passing of the Great Race*, that "it has taken us fifty years to learn that speaking English, wearing good clothes and going to school and to church does not transform a Negro into a white man" (1918, 16).

16. See Pater, *Marius the Epicurean*, where "strenuous self-control and *ascêsis*" is the precondition for aesthetic pleasure, "keep[ing] the eye clear" by "a sort of exquisite personal alacrity and cleanliness" (1985, 49; 54). On the Stoic ascesis, see Foucault, "Technologies of the Self" (1997, 223–52).

17. For readings that emphasize the liberation rather than the control of desire in *Her*, see Laity 1996, 32–42; Du Plessis 1986, 63–69; and Friedman 1990, 105–128.

18. H. D., letter to Winifred Bryher, 14 February 1919, quoted in Gregory 1997, 105. On the psychological conflict between "Western" "wisdom" and "Graeco-Asiatic" "sensuousness," see H. D., "The Wise Sappho" 1982, 63.

19. On Pater's appropriation of Sparta as an example of the "fashioning of oneself into a perfect work of art," see Rawson 1969, 362–63.

20. See Pater, "The Age of Athletic Prizemen," on "the would-be virile Amazon" (*Greek Studies* 1925b, 297).

21. On the admiration of the Greek boy athlete as a late Victorian code for male homosexuality, see Jenkyns 1980, 222–26. Dowling interprets the late Victorian revival of an "idiom of Greek ideality" as an attempt to renew an England "fractured by the effects of laissez-faire capitalism and enervated by the approach of mass democracy," by revealing the martial origins of *paiderastia* (1994, 31, 78). Lears describes a more therapeutic project of "martial antimodernism" in turn-of-the-century America (1981, chap. 3).

22. See Leduc 1994, 235–95; also Cartledge, 1981, 84–105.

23. See Forrest 1975, chap. 10; and Finley 1975, chap. 10. Sparta's decline was attributed by Müller and the "Aryan" historians to the decline in numbers of her ruling class. See Rawson 1969, 336.

24. Quoted in Gregory 1997, 174. The trope was deployed by the American eugenicist and progressivist David Starr Jordan in a lecture delivered at Philadelphia in 1906. Arguing that war is dysgenic since it sacrifices the blood of the best, Jordan quotes from the well-known passage on Greece in Byron's *Don Juan* (iii. 86): "Earth, render back from out thy breast/A remnant of thy Spartan dead!/Of the three hundred grant but three,/To make a new Thermopylae!" (1972, 77).

25. H. D. arrived in England at the moment when the relatively higher fertility of the working class was provoking middle-class fears of "racial degeneration." See Soloway 1990, chaps. 1–4, especially 73–80. According to Soloway, there was, in the Edwardian period, "no end of comparisons with ancient and more recent empires whose flagging reproductive energies explained their decline" (5). On eugenic anxieties about middle-class sterility and working class fertility in Eliot's *The Waste Land* (1922), see Trotter 1984, 42–43; see also his "Modernism and Empire: Reading *The Waste Land*" 1986, 143–53. For a suggestive reading of attempts by modernist writers to forge new "affiliations" consequent on the "failure of the generative impulse," see Said, *World, Text, Critic* (1983, 16–20).

References

Arnold, Matthew. 1885. *Discourses in America*. London: Macmillan.

Bernal, Martin. 1987. *Black Athena: The Afroasiatic Roots of Classical Civilization, Vol. 1: The Fabrication of Ancient Greece 1785–1985*. London: Free Association Books.

Bhabha, Homi K. 1994a. "The Other Question: Stereotype, Discrimination, and the Discourse of Colonialism." In *The Location of Culture*. London: Routledge.

———. 1994b. "Sly Civility. In *The Location of Culture*. London: Routledge.

Bourdieu, Pierre. 1986. *Distinction: A Social Critique of the Judgement of Taste*. Translated by Richard Nice. London and New York: Routledge and Kegan Paul.

Butler, Judith. 1993. *Bodies that Matter: On the Discursive Limits of "Sex."* New York: Routledge.

Cartledge, Paul. 1981. "Spartan Wives: Liberation or Licence?" *Classical Quarterly* 31 no. 1: 84–105.

Darrow, Clarence. 1926. "The Eugenics Cult." *American Mercury* 8 no. 30: 129–37.

Doolittle, Hilda [H. D.] [1920] 1982. "The Wise Sappho." *Notes on Thought and Vision and The Wise Sappho*. Reprint, San Francisco: City Lights.

———. 1984. *Her*. London: Virago Press.

Dowling, Linda. 1994. *Hellinism and Homosexuality in Victorian Oxford*. Ithaca, N.Y.: Cornell University Press.

Du Bois, W. E. B. [1899] 1967. *The Philadelphia Negro*. Reprint, New York: Benjamin Blom.

———. [1903] 1989. *The Souls of Black Folk*. Reprint, New York: Bantam Books.

Du Plessis, Rachel Blau. 1986. *H. D.: The Career of That Struggle*. Bloomington: Indiana University Press.

Dyer, Richard. 1997. *White*. London: Routledge.

Finley, M. I. 1975. *The Use and Abuse of History*. London: Chatto and Windus.

Forrest, W. G. 1968. *A History of Sparta 950–192 BC*. London: Hutchinson.

Foucault, Michel. 1997. "Technologies of the Self." In *The Essential Works of Michel Foucault*. Vol. 1, *Ethics: Subjectivity and Truth*, edited by Paul Rabinow. London: The Penguin Press.

Frederickson, George M. 1987. *The Black Image in the White Mind: The Debate on Afro-American Character and Destiny, 1817–1914*. Hanover, Conn.: Wesleyan University Press.

Friedman, Susan Stanford. 1987. "Modernism of the Scattered Remnant: Race and Politics in H. D.'s Development." In *Feminist Issues in Literary Scholarship*, edited by Shari Benstock. Bloomington: Indiana University Press.

———. 1990. *Penelope's Web: Gender, Modernity, H. D.'s Fiction*. Cambridge, Mass.: Cambridge University Press.

Galton, Francis. 1909. "The Possible Improvement of the Human Breed, Under the Existing Conditions of Law and Settlement." In his *Essays in Eugenics*. London: Eugenics Education Society.

Gilman, Sander L. 1985. *Difference and Pathology: Stereotypes of Sexuality, Race, and Madness*. Ithaca, N.Y.: Cornell University Press.

Grant, Madison. 1918. *The Passing of the Great Race, or, The Racial Basis of European History*, 2nd ed. New York: Charles Scribner's Sons.

Gregory, Eileen. 1997. *H. D. and Hellenism: Classic Lines*. Cambridge, Mass.: Cambridge University Press.

Haller, Mark H. 1963. *Eugenics: Hereditarian Attitudes in American Thought*. New Brunswick, N.J.: Rutgers University Press.

Higham, John. 1963. *Strangers in the Land: Patterns of American Nativism, 1860–1925*. Rev. ed. New York: Atheneum.

Herodotus. 1910. *The History of Herodotus*. Vol. 1. Translated by George Rawlinson. London: J. M. Dent.

Jenkyns, Richard. 1980. *The Victorians and Ancient Greece*. Oxford: Basil Blackwell.

Jordon, David Starr. [1907] 1972. *The Human Harvest: A Study of the Decay of Races Through the Survival of the Unfit*. Reprint, New York: Garland.

Kellogg, Charles Flint. 1967. *NAACP: A History of the National Association for the Advancement of Colored People. Vol 1, 1909–1920*. Baltimore, Md.: Johns Hopkins University Press.

Kevles, David. 1985. *In the Name of Eugenics: Genetics and the Use of Human Heredity*. New York: Alfred A. Knopf.

Klein, Marcus. 1981. *Foreigners: The Making of American Literature 1900–1940*. Chicago, IL.: University of Chicago Press.

Laity, Cassandra. 1996. *H. D. and the Victorian Fin de Siècle: Gender, Modernism, Decadence*. Cambridge, Mass.: Cambridge University Press.

Lears, Jackson. 1981. *No Place of Grace: Antimodernism and the Transformation of American Culture, 1880–1920*. New York: Pantheon.

Leduc, Claudine. 1994. "Marriage in Ancient Greece." In *A History of Women in the West*. Vol. 1 of *From Ancient Goddesses to Christian Saints*, edited by Pauline Schmitt Pantel. Cambridge, Mass.: Harvard University Press.

Lewis, David Levering. 1993. *W. E. B. Du Bois: Biography of a Race 1868–1919*. New York: Henry Holt.

Ludmerer, Kenneth. 1972. *Genetics and American Society: A Historical Appraisal*. Baltimore, Md.: Johns Hopkins University Press.

Michaels, Walter Benn. 1995. *Our America: Nativism, Modernism, and Pluralism*. Durham, N.C.: Duke University Press.

Nielsen, Aldon Lynn. 1988. *Reading Race: White American Poets and the Racial Discourse in the Twentieth Century*. Athens: University of Georgia Press.

Pater, Walter. 1909. "Lacedaemon." In *Plato and Platonism: A Series of Lectures*. London: Macmillan.

———. 1925a. "The Marbles of Aegina." In *Greek Studies: A Series of Essays*. 1895. Reprint, London: Macmillan.

———. 1925b. "The Age of Athletic Prizemen." In *Greek Studies: A Series of Essays*. 1895. Reprint, London: Macmillan.

———. [1885] 1985. *Marius the Epicurean*. Reprint, London: Penguin Books.

———. [1893] 1986. "Winckelmann." Reprint, In *The Renaissance: Studies in Art and Poetry*, edited by Adam Phillips. Oxford: Oxford University Press.

Pickens, Donald K. 1968. *Eugenics and the Progressives*. Nashville, Tenn.: Vanderbilt University Press.

Pound, Ezra. 1962. *Patria Mia and the Treatise on Harmony*. London: Peter Owen.

Rawson, Elizabeth. 1969. *The Spartan Tradition in European Thought*. Oxford: The Clarendon Press.

Ripley, William Z. 1908. "Races in the United States. *Atlantic Monthly* 102: 745–59.

———. [1899] 1923. *The Races of Europe: A Sociological Study*. New York: Appleton and Co.

Said, Edward. 1983. *The World, the Text, and the Critic*. Cambridge, Mass.: Harvard University Press.

———. [1978] 1995. *Orientalism: Western Conceptions of the Orient*. Reprint with a new Afterword. Harmondsworth: Penguin.

The Shorter Oxford English Dictionary. 1977. Edited by C. T. Onions, 3d ed. Oxford: The Clarendon Press.

Solomon, Barbara Miller. 1956. *Ancestors and Immigrants: A Changing New England Tradition*. Cambridge, Mass.: Harvard University Press.

Soloway, Richard. 1990. *Demography and Degeneration: Eugenics and the Declining Birthrate in Twentieth-Century Britain*. Chapel Hill: University of North Carolina Press.

Symonds, John Addington. [1893] 1902. *Studies of the Greek Poets*. Vol. 2, 3d edition. Reprint, London: Adam and Charles Black.

Trotter, David. 1984. *The Making of the Reader: Language and Subjectivity in Modern American, English, and Irish Poetry*. London: Macmillan.

———. 1986. "Modernism and Empire: Reading *The Waste Land*." *Critical Quarterly* 28 no. 1 and no. 2:143–53.

Walling, William English. 1908. "The Race War in the North." *Independent* 65 (September 3): 529–34.

Warner, Sam Bass. 1968. *The Private City: Philadelphia in Three Periods of Its Growth*. Philadelphia: University of Pennsylvania Press.

Re-examining the Political Left: Erskine Caldwell and the Doctrine of Eugenics

Sarah C. Holmes

ERSKINE CALDWELL HAS LONG BEEN CONSIDERED A LEFTIST WRITER, in his own eyes and in the eyes of critics. For example, in a 1931 *New Republic* article, T. K. Whipple wrote that Caldwell's work was indicative of "picaresque, proletarian, peasant America" (Cook 1981, 364). Such critical estimation was warranted. In 1932 the novelist officially supported (with fifty-two other intellectuals) the presidential candidacy of Communist Party nominee William Foster. Caldwell also published work in the well-known leftist publication *New Masses,* such as his 1932 review of Edward Dahlberg's *From Flushing to Calvary* in which Caldwell makes clear his beliefs in radical and Communist ideas (Cook 1981, 362–69). Finally, Caldwell was a contributor to the 1935 anthology, *Proletarian Literature in the United States,* along with other notable leftist writers such as Granville Hicks and Michael Gold.

Readers, especially literary critics, bring certain assumptions to the designation "leftist writer," and those assumptions determine the ways they read a text. Once labeled "leftist," a writer's work is deemed liberal or "subversive" not only in regard to class, but also in regard to gender, race, and sexuality. A reader assumes that a leftist or radical writer positions his or her characters against conservative ideologies of the time. For example, in a frequently cited article entitled "Enough Good Reasons for Reading, Studying, and Teaching Erskine Caldwell," Scott MacDonald writes that Caldwell—a writer with an "awareness of immense social problems" and one who wanted to "spur efforts to ameliorate these problems"—had attitudes about sex that were "quite liberated, particularly with regard to female sex roles" (1979, 15–16). However, such an interpretation presupposes that categories like class, race, gender, and sexuality are somehow separate from the culture in which they exist, as if a work of fiction or nonfiction can step *outside* the very culture in which it resides and impart a "truth" that only a "leftist" writer can purport. MacDonald assumes that because Caldwell is leftist and because his female characters share with their male counterparts the joys of sex, that such a gender characterization is liberating.

When, on the other hand, one examines Caldwell's novel *within* its cultural and historical context, and when one becomes aware of Caldwell's interest in eugenics—including a report stating that he was a member of the American Eugenics Society—one can no longer *assume* a liberal reading of this text (American 2002). In 1930s America—a period in which eugenics was rampant in the mind of the general public, and in the mind of Erskine Caldwell—a poor woman's sexuality was not something liberating but rather was something *degenerate.* A lascivious poor woman was seen as a threat to the genetic makeup of society. Because a writer is leftist, then, that political affiliation does not guarantee that he or she has created a subversive text in regard to gender. In addition, the affiliation does not guarantee that the writer has even fulfilled a duty to the working class. Contrary to popular belief, which assumes that leftist literature is on the side of the working class, a good deal of leftist fiction written during the Great Depression was not working from outside the conservative (mainstream) right but was working *within* it—and therefore was not a critique of eugenics but was *complicit* with that ideology. It is critical, therefore, to reexamine a writer like Caldwell and to rethink his work in light of his involvement with eugenics. For example, when teaching a novel like *Tobacco Road,* faculty should warn students that sexualized female characters are not always "liberating" because such a characterization can carry a tremendous amount of cultural baggage. Likewise, a novel written about poor people may very well reproduce many of the stereotypes that circulate in society. Thus, this essay argues that even leftist writers need to be examined within the confines of their cultural time.

While many critics still consider Erskine Caldwell a leftist, he never adequately fostered sympathy for his poverty-stricken characters. Perhaps the greatest proof that Caldwell was influenced by mainstream culture—in which people did not sympathize with the poor but rather accepted the notion of eugenics—is the original *New Masses* review of *Tobacco Road*, in which Jack Conroy points to the lustful nature of the Lesters: "They are all dying of pellagra and starvation, yet other organs beside their stomachs seem to plague them the most." Conroy also remarks that Caldwell "lacks social understanding" (Cook 1981, 364). It seems clear that some leftist critics doubted Caldwell's "leftist" politics due to his disparaging portrayals of the Southern poor. Erskine Caldwell himself acknowledged that there were inherent difficulties with writing leftist literature in the 1930s. For instance, in 1932 he wrote, "We cannot expect to write or read genuinely proletarian novels until we live in a proletarian world. At our present stage we find that we can only begin where our previous existence dumped us. We were dumped by a capitalist system on hard ground, and here we lie. . . . We are scattered, broken, and bewildered" (Cook 1981, 365). Here Caldwell suggests a lack of faith in adequately representing the proletariat. While the popular press has continued to categorize Caldwell as a leftist, it is quite clear that novels like *Tobacco Road* are steeped in other ideologies that circulated in 1930s America. Specifically, the representation of poverty in *Tobacco Road* reproduces a eugenic conservatism—the notion that moral and intellectual deficiencies are genetically inherited. Caldwell subverts sympathy for the poor through his eugenic discourse of landscape, metaphor, and characterizations.

Because Erskine Caldwell and his readers advocated eugenics, *Tobacco Road* worked dialogically with the eugenic discourse that circulated in 1930s America. As Mikhail Bakhtin writes of the processes of dialogism, "the dialogic orientation is obviously a characteristic phenomenon of all discourse. It is the natural aim of all living discourse. Discourse comes upon the discourse of the other on all the roads that lead to its object, and it cannot but enter into intense and lively interaction with it" (Todorov 1984, 62). Caldwell's novel enters into an on-going dialogue with his father's eugenic study, with other eugenic family studies, and with its readers' perceptions of the poor. There is an ideological conversation among these texts, readers, and institutions. Tzvetan Todorov refers to this conversation as "intertextuality" and writes that an utterance, a speech act that proposes a cultural position, "enters in relation with past utterances that had the same object, and with those of the future, which it foresees as answers" (Todorov 1984, 53). *Tobacco Road* converses with other eugenic utterances in the larger cultural landscape.

Tobacco Road's Source

Erskine Caldwell wrote about one of the most impoverished regions in American society and based his novel on his father's eugenic study of a Georgian family (cf. Keely). One of the most popular trends in the eugenics movement was the examination of such "unfit" families. Eugenicists tallied up how many criminal, feebleminded, insane, immoral, and unintelligent people constituted one lineage and "proved" that these traits were genetic. Ira Caldwell, a preacher and a journalist for the *Augusta Chronicle*, published his findings about the Bungler family in a serialized article in the magazine *Eugenics: A Journal of Race Betterment*. Not surprisingly, the senior Caldwell concluded that the Bunglers were inherently feebleminded, lazy, and—as would be most important to readers during the New Deal administration—beyond help (Miller 1995, 124).

In 1929, Ira Caldwell began a social experiment in hopes of helping the tenant farmers of Georgia; his goal was to rehabilitate a family by changing their environment. He encountered the Bunglers, a family "marked by illiteracy, incest, feeblemindedness, [and] hookworm" (Miller 1995, 124). Caldwell arranged for them to leave Burke county and move into Wrens county, provided them with schooling and work, fed them, and found them a church to attend. Within a few months, however, Ira realized that his assistance was not helping this family to escape the bounds of poverty and degeneracy. The Bungler children stopped going to school, the father quit his job, and the family moved back to their barren farm and house. After his help proved fruitless, he investigated their ancestry and concluded that the Bunglers were eugenically unfit. He then wrote a five-part article in which he concluded that people like the Bunglers should be sterilized. Throughout the article, Ira Caldwell incorporates a series of eugenic discourses that are similar to the rhetorical strategies of *Tobacco Road*. Specifically, he discusses the recklessness of the Bunglers, traces their degeneracy through genetics, equates these individuals with animals, utilizes epidemic terminology to describe them, marginalizes them from the rest of society, adopts the role of ethnographer, and argues that society should stop the Bunglers from reproducing (a program known as "negative eugenics").

Caldwell spends considerable time discussing the recklessness of the Bunglers. For example, the Bungler boys were sent to an army recruiter, failed the intelligence tests, and subsequently were unable to join the military. The boys then worked in the mills with their father, and they made more money than they had ever made before. (In preceding years, they had trouble making enough to eat.) Once the father and sons had money, however, they "did not know how to spend wisely" and squandered it "in an effort to satisfy common desires" (I. Caldwell 1930, 205). Caldwell also informs us that the Bunglers would spend so much money that they would run out of food by midweek so that the management at the mill had to give them a loan to tide them over. Caldwell examines the Bunglers' careless behavior and traces this trait to their genetics. After conducting an elaborate study of the Bungler heritage, Ira explains that certain characteristics are inherent in most of the family members. Caldwell thereby details the genetic determinism, recklessness, and lack of intelligence in the Bungler family line as a means to support his eugenic argument.

After asserting the genetic degeneracy of the Bunglers, Caldwell's strategy is to differentiate the Bunglers from the rest of society. His comparison between animals and eugenically unfit people is one effective method. By first establishing their subjects as mindless *animals,* eugenicists thereby stressed corporeality and sexual license over an ability to think rationally. This was particularly true for the Bunglers, to whom Caldwell attributed a "reproduction impulse" (209). In other words, it is easier to persuade a reader that people should be sterilized if they breed "[l]ike bees" (209). If they are animals, they can no longer rise above their class—because they are without reason. Once class is accepted as a *natural* classification, as opposed to one borne out of a capitalist system, eugenicists can rationalize class hierarchies and weed out those members of the country who are, in their minds, draining the economy. The natural classification is established by the Bunglers' association with animals, an association that was the cornerstone of eugenic reasoning. America's leading eugenicists argued that if specialists breed animals, they should also be able to breed out unfit human beings—and that our society depended on it if we were to promote the human race. Charles Davenport, the father of American eugenics, remarked that "the most progressive revolution in history" could be achieved if "human matings could be placed upon the same high plane as that of horse breeding" (Kevles 1985, 48). Similarly, Ira Caldwell suggests that society should weed out people like the Bunglers, who he claims have "dwarfed intelligences and lean souls" (1930, 377).

Caldwell uses the analogy of birds to describe the Bunglers, a recurring metaphor that reinforces his suggestion that misfit citizens need to be sterilized:

> There is a certain bird that hides in the dark shadows of the forest. It is quiet and makes only short flights close to the earth. Many people have never seen these birds; do not know that they exist. Likewise there are many Bunglers and their like hidden away in the hinterland of the country districts. They are in the mountain coves and they are on the unpaved alleys in the cities. Many of these people in the country are never seen because they are never on parade. They have nothing to parade but squalor. In the city they are lost in the crowd and the cry of their crushed spirits is lost in the roar of the mill. But they are here and they are everywhere. If the powers that be in this country are wise, a period of social and economic reconstruction such as the world has never seen will be undertaken at once. (336)

In this passage, Caldwell achieves two goals at once through some fancy rhetorical footwork. He marginalizes the Bunglers, yet also gives them the power to infect the center of society. In one respect, the Bunglers are "hidden away," but Caldwell simultaneously imagines them as being everywhere. Like an epidemic, the Bunglers (and others like them) taint the pure landscape of America. They are like germs—nowhere and everywhere—and they are insignificant and catastrophic. The Bunglers are distant enough to constitute a colony, and yet close enough to stir fear. Two things result from such a comparison. First, the Bunglers and other families become viable subjects for scientific study. Second, fears about the poor are diminished for the general public—because questionable genes will not muddy the gene pool if America acts quickly. It is Caldwell's opinion that these types of people should be studied since they are silently infiltrating not just the margins, but the whole of society. Such a passage, therefore, supports his eugenic call for action: "Self interest as well as the finer calls of humanity demand a lessening of the swelling tide of inefficiency" (1930, 204).

Caldwell continues to marginalize the Bunglers by placing them into a separate geographical and cultural space from the rest of society. A type of metaphoric colonialism takes place in which geographical separation augments their otherness. Ira Caldwell plays ethnographer here, constantly distancing himself and the rest of the country from the Bunglers as if he is the Westerner studying the third-world "other." He even suggests that the Bunglers, although residing in Georgia, are part of an un-American culture: "The tendency to withdraw from the world as far as possible, to live their own lives as independently as humanly possible has resulted in building up a sociological Chinese wall that shuts the Bunglers in and shuts the world out. . . . one of the most important factors in the development of a civilization is the cross-fertilization of ideas. For obvious reasons, there is very little cross-fertilization of this type in this family" (247). Caldwell puts himself in the position of a civilized scientist who must study and illuminate the ways of the uncivilized. He also preys upon Depression-era readers' fears about race and "otherness" by racializing the Bunglers as Chinese—as *foreign*.

As part of this marginalization, Caldwell imagines the Bunglers occupying a space that is similar to how America once saw its frontier: "One of the outstanding characteristics of the Bungler family is the tendency to live in sequestered places. From social timidity or from economic necessity they have a well defined tendency to live in out-of-the-way localities. Of the many Bungler homes visited in the prosecution of this survey not one family was found living on main roads through the country; indeed, a comparatively few were found on public roads of any type. They apparently prefer to live on blind roads that lead nowhere, their homes often being at the end of the trail" (247). Caldwell recreates an American frontier where the Bunglers live. They are at the "end of the trail," an allusion to an earlier frontier rhetoric that once had great implications in this nation's history. The frontier was a place to be both harnessed and harvested. Similarly, by adopting this rhetoric, Caldwell gives more weight to his later discussion of eugenic intervention. If a family like the Bunglers chooses to live on blind roads, and if they encompass a frontier that has yet to be discovered, then it is up to the eugenic explorers to lead us into Bungler territory and civilize it. Caldwell often uses the discourse of expansion when discussing the Bunglers, as when he suggests that their whereabouts in other states "constitute[s] an unexplored region for eugenical research" (209). In addition, he merges this frontier rhetoric with his ethnographic rhetoric: "To all practical purposes these people live in little kingdoms with closed frontiers. Their little communities are social Thibets into which visitors seldom go and out of which messengers seldom come" (249). He culturally distances the Bunglers and instills a sense of danger, thereby suggesting that society puts itself at risk the longer it "keep[s] the Bunglers and their type hidden away in the cellar of sub-consciousness" (334).

One factor, however, made it impossible for society to store the Bunglers away in their subconscious; they, and other people like them, kept having children. Caldwell addresses their virility, which was a great concern for eugenic advocates. He claims that Thomas Bungler had twelve children by one wife, remarried, and eventually became the grandfather of eighty-five children and the great-grandfather of forty-five children (all while he was still alive). Caldwell estimates that there were "three thousand descendants of the original Georgia Bungler" (210). He compares them to bees in an effort to express their propensity for reproduction: "Like bees, his [Thomas Bungler's] descendants have swarmed into adjoining counties and neighboring states. The reproduction impulse of their ancestors persists in almost all of the present generation, and hence the family tree is very large" (209). He argues that Benjamin Bungler's father, Thomas Bungler, should have received the "eugenic advice" of birth control because of the rapid growth of his family (209).

The problem is that the Bunglers were the *wrong* type of family to procreate:

> What is taking place in the United States today? The Bunglers are producing very large families. The college-bred people are rearing small families. On a certain corner in New York, recently, a young woman thrust birth control literature into the hands of all who passed. The same afternoon the author went to a movie on the Bowery. A trip through the east side showed children as thick as leaves in Vallambrosa. But there was no birth control literature given away in that section. The Bunglers cannot read birth control literature and what passes for the intelligentsia does not need it. We need birth control but it is now effective at the wrong end of society. (I. Caldwell 1930, 380–81)

Caldwell quotes a woman in the Bungler family who admitted their plight: "the Bunglers are decreasing in every way except numbers" (250). More than the lack of birth control, the Bunglers are also "breeders," and Caldwell quotes one family member who says his latest wife "breeds too fast" (293). Caldwell warns his readers: "The Bunglers are here and they are multiplying" (381). He also writes that "[a]t the present time there are 1,000,000 people in the United States who are, to all practical purposes, idiots" and that these people "are the greatest perils of a republic" (381). In a letter during the 1930s, Ira suggested to Erskine that he write a book about the middle class who were dying out because of the increasing numbers in the lower class (Caldwell Papers).

Ira Caldwell suggests that the Bunglers be sterilized. He

writes, "There is only one feasible conclusion and that conclusion is that we have hidden away in the social order, a vast number of people who are a source of weakness rather than of strength" (335). He places the Bunglers in a larger arena, one that places their existence at the core of national trauma: "What of the future in a larger sense? What do the Bunglers and thousands of others of their type mean to civilization? What is the trend of the social order at the present time?" (378). Ira Caldwell concludes his essay with a eugenic call to action that will solve America's problem: "Sterilization of Jukes, and Nams, and Bunglers would lessen the pressure from the lower levels of society" (383). Caldwell puts the Bunglers in the same category of the undesirable families whom eugenicists studied. His son's fictional family, "The Lesters," would join this assemblage of undesirables.

Tobacco Road

In *Tobacco Road,* Erskine Caldwell appropriates his father's political ideas *and* rhetorical strategies. The younger Caldwell reveals his eugenic subtext—that people like the Lesters should be prevented from having children. Specifically, Erskine Caldwell utilizes character development, marginalization, racialization, comparisons between people and animals, an argument for genetic determinacy, a description of the poor's recklessness, and an ethnographic element.

The main impetus for Erskine Caldwell creating the Lesters was the Bunglers. Both poor and Georgian, the family structures are similarly developed. The marriage of Sister Bessie and sixteen-year-old Dude is strikingly similar to a description of a couple's marriage in Ira Caldwell's Georgia: "the woman pastor of the church . . . was the widow of a Methodist minister. . . . Subsequently she began to preach and was the pastor of the church of rattlesnake fame for a number of years. At the age of fifty odd years she married a school boy about sixteen years old. Although the boy had not finished the fourth grade in school he alternated with his bride in preaching" (I. Caldwell 1930, 334). Besides the obvious comparison to be made between the characterization of Bessie and this woman preacher (their age, vocation, and a sixteen-year-old second husband), Bessie seems to be fashioned out of this real woman's eccentricity and her desire to have her young husband preach with her. Likewise, Caldwell took the name Dude Lester from his father's study, specifically Dude

Bungler. Later Dude Bungler would also become one of the subjects of the photographic essay book that Erskine Caldwell produced with his wife, Margaret Bourke-White. Caldwell also used the Bunglers in a series of articles in the *New York Post* (Klevar 1993,178–80).

As interesting as these surface-level similarities are, there are significant rhetorical similarities between the two Caldwells' works. One of the most striking aspects of *Tobacco Road,* for instance, is the recurring marginalization of the Lester family from the rest of society. All of their extended family live elsewhere; everyone of social status lives in outlying counties; and the Negroes only pass by their house occasionally. Both Ira and Erskine Caldwell spend a great deal of time relegating their subjects into separate cultural spheres from the rest of society. While Ira's Bunglers inhabit "the frontier," Erskine geographically distances the Lesters from the community, a symbolic manifestation of their "otherness." Their spatial distancing creates a metaphor for their status as an abject part of society—one that can be thought of as separate and disposable. One passage in particular is worth quoting in order to illustrate Caldwell's rhetorical strategy:

> After seventy-five years the tobacco road still remained, and while in many places it was beginning to show signs of washing away, its depressions and hollows had made a permanent contour that would remain as long as the sand hills. There were scores of tobacco roads on the western side of the Savannah Valley, some only a mile or so long, others extending as far back as twenty-five or thirty miles into the foothills of the Piedmont. Any one walking cross-country would more than likely find as many as six or eight in a day's hike. The region, topographically, was like a palm leaf; the Savannah was the stem, large at the bottom and gradually spreading out into veins at the top. On the side of the valley the creeks ran down like the depressions in the palm leaf, while between them lay the ridges of sand hills, like seams, and on the crests of the ridges were the tobacco roads. (E. Caldwell 1995, 64)

Caldwell portrays the fears that were filtering through American culture in 1932, namely the fear that the rural poor were making an indelible mark on the social "landscape" of America. Moreover, the tobacco road is literally on the margins of the countryside, on its "frontier." Similar to Ira Caldwell's work, this discussion of the frontier is marked by serious class and eugenic inflections. When Erskine Caldwell describes the tobacco road's depressions and hollows, one understands that only *certain* classes inhabit the "hollows" of America. At the same time, he compares their homeland to a leaf—a metaphor that implies their heritage can at any time blow away in the winds of cultural change.

Erskine Caldwell also differentiates the Lesters from the rest of society because they inhabit a different kind of figurative space. The family is either caged or inhabiting a stage, as if they are performing a freak show for an audience because they are drastically different from others. Benjamin Reiss refers to early freak shows as "spectacles of race," whereby whites enjoyed watching the spectacles of blacks (1999, 78). Caldwell, however, reverses the self/other distinction upon which the freak show was based. Consistent with racist ideology of the period, then, Caldwell deprecates the Lesters by suggesting they are inferior to black people. For example, the black characters avoid interaction with the Lesters, implying that they look down on the odd family:

> Dude hollered at them, calling their names; but none of them spoke. They stopped and watched.
>
> "Howdy, Captain Lov," one of them said.
>
> Lov did not hear. The Lesters paid no more attention to the negroes. Negroes passing the house were in the habit of looking at the Lesters, but very few of them ever had anything to say. Among themselves they talked about the Lesters, and laughed about them; they spoke to other white people, stopping at their houses to talk. (E. Caldwell 1995, 28)

Here, black people are on the same social level as other white people, with the Lesters being inferior to all of them. In a widely racialized culture in which blacks were labeled as degenerate because of their putative sexual prowess and inferior intelligence, it is significant that Caldwell characterizes the Lesters as even beneath blacks. The black people will talk to other whites, but they will only *laugh* at the Lesters. The Lesters are metaphorically on stage again, performing a freak show for the blacks. Because the Lesters are lower on the social scale than the blacks, they are lower than other (fit) whites. Race proves to be a vehicle by which Caldwell groups the Lesters into the lowest human rung on the evolutionary ladder. The result is to segregate the Lesters from "good" society.

Caldwell uses this segregation to instill fear in his other characters. For example, Lov comes to visit, and he is afraid of the Lesters, who metaphorically seem to be caged in their yard (to protect those on the *outside*). Lov tries to avoid their space altogether: "Usually when he came by the Lester place with turnips or sweet potatoes, or for that matter with any kind of food, he left the road half a mile from the house and made a wide circle through the fields, returning to the road a safe distance beyond. To-day, though, he had to speak to Jeeter about something of great importance, and he had ventured closer to the house than he had ever done before. . . . He was debating within himself the danger of entering the yard, against the safety of staying where he was in the road" (E. Caldwell 1995, 1–2, 5). Coupled with the fact that Ellie May and the grandmother act like wild animals, such as when Old Mother Lester crawls out from under the front porch, Caldwell metaphorically cages his characters and further marginalizes them from the rest of society.

Caldwell also substantiates the inferiority of his characters by rendering them as animals and, by association, as libidinous and primitive. For example, Dude remarks to Jeeter, "Ellie May's acting like your old hound used to do when he got the itch" (18). This passage is charged with a stereotype about lower-class women, likening them to sexualized animals who cannot control their urges. Here, Caldwell makes it clear that class, gender, and sexuality are linked. Also, Caldwell makes a Cartesian distinction between a rational, upright human and someone like Dude, a soulless four-legged animal who cannot pray properly: "They knelt down to pray. Dude got down on all fours, looking straight into Bessie's nose while her eyes were closed" (104). As the black people walk by and laugh at the Lesters' animalistic antics, the family moves further down the evolutionary chain.

Caldwell alludes to genetics as a reason for why the Lesters are degenerate. All of the Lesters are ignorant, disfigured, or lustful people, with the exception of Pearl. He describes her as being far different from the rest, a pearl in the rough: "Pearl's long yellow curls hanging down her back, and her pale blue eyes, turned Lov's head. . . . It would have been impossible for her to dress herself, or even to disfigure herself, in a way that would make her plain or ordinary-looking" (25). Traditionally, the name Pearl has represented purity, and Caldwell provides a reason for her purity: "Dude did not have very much sense, and neither did one or two of the other children, and it was natural for him [Lov] to think that Pearl was afflicted in the same way. The truth was, Pearl had far more sense than any of the Lesters; and that, like her hair and eyes, had been inherited from her father. The man who was her father had passed through the country one day, and had never been seen since. He had told Ada that he came from Carolina and was on his way to Texas, and that was all she knew about him" (31). Pearl comes from a genetic line that is different from the rest of the Lesters, and hence she is prettier and smarter. The Lesters' degeneracy and deformities, then, are genetically inherited traits. Their biology may be one reason why Pearl refuses to consummate her marriage with Lov; she will not destroy the genetic superiority of her father's family line.

Finally, one of the most convincing ways that *Tobacco Road* is a eugenic tract is its main plot device: Bessie's appropriation of a new car. Bessie buys her car because of the money she receives from her husband's death, and this money is a metaphor for public assistance (i.e., the dole) that poor people received under New Deal programs. *Tobacco Road* struck a cord with Depression-era readers because of its attention to such public welfare. Many people had opposed the dole, especially because of the genetic theories espoused by eugenicists. At an exposition in Philadelphia, the American Eugenics Society erected a billboard that announced with flashing lights that every fifteen seconds an American citizen's one hundred dollars went towards the care of people with bad heredity and that in the United States a mentally deficient person was born every forty-eight seconds (Wiggam 1927, 143). The American Eugenics Society also published *Tomorrow's Children* in 1935, in which Ellsworth Huntington, the president of the society, directly addressed the issue of federal aid and informed the reader that financial aid alone would not suffice for someone who was naturally deficient. America was very concerned with how much money it cost to help the poor—whom they considered unfit.

The interest in eugenics and the concern about the dole both had serious repercussions, such as the growth of sterilization as a strategy to weed out the unfavorable. By the end of the 1920s, twenty-four states had sterilization laws to curb America's spending on unfit children, and the popularity of sterilization grew dramatically during the Depression as America struggled with an economic crisis. By the middle of the 1930s, approximately twenty thousand legal sterilizations had been performed in this country. People supported sterilization as a punishment for those who were on public relief past a certain length of time (Kevles 1985, 111–14). A former member of the Virginia Board of Supervisors remarked on the economic aspect of sterilization:

"They were hiding all through these mountains, and the sheriff and his men had to go up after them. . . . They really got them up on Brush Mountain. The sheriff went up there and loaded all of them in a couple of cars and ran them down to Staunton [Western State Hospital] so they could sterilize them. . . . People as a whole were very much in favor of what was going on. They couldn't see more people coming into the world to get on the welfare" (Kevles 1985, 116). The eugenics movement was driven by a fear of giving free aid to the poor, who were considered to be intellectually defective and too unfit to make proper use of it. In a 1936 article entitled "The Survival of the Unfittest," Channing Pollock summarized the public's feelings when he defined the dole as "Government's corruption of its citizens" (38). He also asked this question: "Who is to pay all this? the fit, of course" (39).

This question about public relief is a crucial point in our analysis of Caldwell's novel because the main plot event is the poor's appropriation of free money. As the novel unfolds, we see that the Caldwell characters are simply too unfit to use this money wisely—again, an important theme in many eugenic tracts. Bessie, a garrulous widow, is at the center of Caldwell's critique of public aid. She is freakish, ignorant, and disfigured. As is often the case in literature, Bessie's deformed nose symbolizes a mental deficiency. Such a deformity is also a clear sign for eugenicists in identifying those who are not fit. In other words, her appearance supports the characterization of her as an inferior and ignorant person. Therefore, Caldwell intersperses physical descriptions of Bessie with descriptions of her recklessness. Even though the Lesters are starving, she purchases a new car rather than helping them. She remarks about the eight hundred dollars and the new car she will buy with it: "I aimed to use it in carrying on the prayer and preaching my former husband used to like so much. I always did want a brand-new automobile" (E. Caldwell 1995, 83). Her plan is to marry sixteen-year-old Dude so he can drive her to various places where they will preach. Given the tumultuous climate at the time and the American public's resistance to giving money to undeserving people, Caldwell's message is very clear: *the poor will not spend the money wisely*.

In *Tobacco Road*, we see what happens when the "unfit" obtain free money. For example, the car dealer takes advantage of Bessie. Instead of inquiring about the price, she foolishly tells the salesman how much money she has. After winking to his partner, the salesman tells her that she can have the Ford for eight hundred dollars and drive away with it that same day. He lies about the registration laws and then sells her the car for all the money she has. Another indication that Dude and Bessie are too "unfit" to spend money wisely is that they total the car within days of receiving it. During the first few days, Dude and Bessie drive with an exaggerated reckless abandon: "Dude was driving about twenty miles an hour, and he was so busy blowing the horn he forgot to slow down when he turned into the yard. The car jolted across the ditch, throwing Bessie against the top three or four times in quick succession, and breaking several leaves of the rear spring. Dude slowed down then, and the automobile rolled across the yard and came to a stop by the side of the house" (E. Caldwell 1995, 100). This passage is significant for a number of reasons. First, Caldwell makes clear in the previous chapter what simple pleasure Dude gets from hitting the horn, as if he is a child. Second, Dude is so careless that he potentially injures his new wife. Third, Dude and Bessie ruin their new purchase—an indication that poor people are unable to use money wisely or take adequate care of their possessions.

Within a span of days, Dude and Bessie wreck the car. They enter a filling station for some assistance, and the owner remarks, "You've already ruined your new car" and "That's a shame. I hate to see people who don't know no better ruining automobiles" (139). The car is ravaged, but Bessie and Dude remain indifferent: "After their first accident, when Dude ran into the back end of the two-horse wagon near McCoy and killed the colored man, anything else that happened to the car would not matter so very much, anyway" (153–54). Bessie and Dude do not have moral or intellectual sensibilities that would render them responsible human beings. Sylvia Jenkins Cook remarks that Dude's and Bessie's treatment of the car is "a comment not only on the idiocy and profligacy of Dude, but more generally on the warped priorities of poor people" (366). This interpretation of the "unfit" is further accentuated when Dude runs over his grandmother and leaves her lying in the dust without any attempt to help her: "He saw the old grandmother lying in the yard and he slowed down to look at her, but he did not linger there" (166). Caldwell makes it clear that poor people are dumb, irresponsible, and immoral. Simply put, Caldwell asks: *why give poor people public assistance if this is how they act?*

Tobacco Road's Influence

Because of the eugenic source and subtext in *Tobacco Road,* Caldwell's influence on American cultural ideology was substantial during the Depression. Specifically, Caldwell described the supposed degeneracy of his characters by utilizing the rhetorical voice of ethnography, as Ira Caldwell did in his Bungler study, and in doing so created a novel that readers saw as a realistic and truthful account about a particular population—the Southern poor. Even when discussing his work, Caldwell adopted the persona of a sociological ethnographer when he said that all he wanted to do was to describe the lives of Southern poor whites (Brinkmeyer 1979, 47). Caldwell considered his father's sociological work to be a great influence: "Well, as far as my father was concerned, I don't think he was shocked by anything I wrote because he knew more about life than I did. He was more of a sociologist than he was anything else, and he was very familiar with the lowest rung of existence, so that I could have learned from him, and I did learn from him, very much" (Kelly and Pankake 1984, 38–39). In *Call it Experience,* Erskine writes that he "wanted to tell the story" of people as "they actually lived their lives from day to day and year to year without regard for fashions in writing and traditional plots" (Brinkmeyer 1979, 47). Caldwell achieved this goal, because readers did indeed accept the novel as a piece of ethnography.

Scholars have acknowledged the ethnographic aspects of Caldwell's novels. In the introduction to *Tobacco Road,* Lewis Nordan writes in regard to both *Tobacco Road* and *God's Little Acre,* "In important ways the books were indeed sociological," and the sociological nature of the books served to strengthen class bias by creating a we/them, self/other distinction (Nordan 1995, vi). Nordan also writes, "When we accused others of living on Tobacco Road, we were distancing ourselves from a sociological stratum of society that we were afraid of being associated with, for that is how we—or I, anyway—understood *Tobacco Road,* as a sociological statement about a region" (vi). Similarly, Robert Brinkmeyer writes, "Everyone accepts that Caldwell wrote the novel with a social purpose in mind, namely to draw attention to the plight of the tenant farmers and to make some suggestions on how to improve their condition" (Brinkmeyer 1979, 47). *Tobacco Road,* then, works like sociological ethnography because of the combination of Erskine Caldwell's stated intentions and the readers' responses to this book—readers who assumed that they were learning the facts about this unusual and freakish community. Thus, the novel works less on a fictional plane than it does on a journalistic or documentary plane, thereby contributing to America's stereotypes about the poor.

Given Caldwell's readership—the American public during the

1930s—the novel helped fuel the fear and disgust that the "other" provoked in people, particularly in those who influenced social policies during the Depression. In an editorial in the *Augusta Chronicle* in 1935, a social worker appropriates Caldwell's novel for his or her own political objective:

> Having been a social worker for many years, and dealing with the class of people about whom Mr. Caldwell writes, I submit a suggestion. These people can not be lifted to any standard of normal life as long as they are permitted to control themselves. There should be a place, not a prison, but some kind of institution where they may be placed under the law, and forced to stay. . . . These people are not necessarily natives of Georgia, but because of the mild climate here they drift to the South from every other section, and they can find miserable huts all over the state, which they can crawl into, and find wood to burn, so they do not freeze to death as they would do in other parts of the country. I also suggest a campaign to burn or tear down all the wretched old huts and shacks. . . . There is no use to be sentimental about the matter. If a patient has a cancer it is cut out. These people are a cancer on society, a menace to themselves and the state; and to perpetuate the condition only increases their number. There should be a law to control the situation and a remedy to back up the law. (L. Holmes 1981, 134)

In this passage we see the effects of *Tobacco Road* on a Depression-era reader. The social worker expresses contempt for Caldwell's characters as representatives of *all* the poor and unfit people who burden society. Also, Ira Caldwell wrote articles for the same newspaper in which this editorial appeared (Miller 1995, 124). Most readers, such as this social worker, would have been familiar with both Ira's and Erskine's writings, and therefore would not differentiate between the two as a contemporary reader would. Therefore, a "leftist" novel like *Tobacco Road* worked intertextually with eugenic tracts during the 1930s. Both works—"The Bunglers" and *Tobacco Road*—shared similar results whereby their readers presumed that poor people needed to be exterminated from society.

While one may assume that Ira Caldwell's study is the "eugenic" text and Erskine Caldwell's novel is the "literary" text, both of these works are *eugenic* texts. Given the cultural climate in the 1930s, an ideology like eugenics produced these works and their popular reception. And because of the novel's popularity, *Tobacco Road* reached an audience that "The Bunglers" would have missed. The novel sold millions of copies, was translated into different languages, was made into a successful play that ran sixteen hundred performances, became a movie directed by John Ford, and was one of the reasons why William Faulkner considered Caldwell to be one of the best writers of the time (Nordan 1995, vi). Because of its popularity, the novel had a great impact on American culture. As Louis Nordan suggests, people confused poverty with moral degeneracy, and Caldwell's novel helped to implant that idea into the minds of the American public: "[W]hen we spoke of the poorest, or the most hopeless, or even the morally reprehensible among us, we said, 'They might as well be living on Tobacco Road.' I had never heard the name of Erskine Caldwell, let alone read one of his books; yet these words, and this vision of the rural South, had made their way into the American mind and into our vernacular" (v). Because of such a response, then, the novel cannot be separated from its politics. Caldwell's work fits Mikhail Bakhtin's definition of a novel: it is a genre that "crosses the boundary of what we strictly call fictional literature" (Bakhtin 1981, 33). The novel played a crucial role in the eugenic discourse of 1930s America as it propagated eugenic ideas about the unfit—ideas that would soon contribute to America's indifference to Nazi eugenic policies. In fact, *Tobacco Road* may be one of the most successful eugenic tracts in American history and one that puts into serious question the ideologies and goals of the political left.

References

American Eugenics Society Membership Activities. 2002. In *Eugenics Watch.* Available from www.africa2000.com/ENDX/aemema.htm.

Bakhtin, Mikhail. 1981. *The Dialogic Imagination: Four Essays*, edited by Michael Holquist. Austin: University of Texas Press.

Brinkmeyer, Robert H., Jr. 1979. "Is that You in the Mirror, Jeeter? The Reader and *Tobacco Road.*" *Pembroke Magazine* 11:47–50.

Caldwell, Erskine. [1932] 1995. *Tobacco Road.* Athens: University of Georgia Press.

———. Papers. Syracuse University Library, New York.

Caldwell, I. S. 1930. "The Bunglers." Parts 1–5. *Eugenics: A Journal of Race Betterment* 3, no. 6:203–10; no. 7:247–51; no. 8:293–99; no. 9:332–36; no. 10:377–83.

Cook, Sylvia Jenkins. [1979] 1981. "Erskine Caldwell and the Literary Left Wing." Reprint, in *Critical Essays on Erskine Caldwell*, edited by Scott MacDonald. Boston: Hall and Co.

Holmes, L. E. [1935] 1981. "From a Social Worker." Reprint, in *Critical Essays on Erskine Caldwell*, edited by Scott MacDonald. Boston: Hall and Co.

Holmes, Sarah C. 2002. "Leftist Literature and the Ideology of Eugenics in the American Depression." Ph.D. diss., University of Rhode Island.

Huntington, Ellsworth. 1935. *Tomorrow's Children*. New York: Wiley Press.

Keely, Karen A. 2002. "Poverty, Sterilization, and Eugenics in Erskine Caldwell's *Tobacco Road*." *Journal of American Studies* 36, no. 1: 23–42.

Kelly, Richard, and Marcia Pankake. 1984. "Fifty Years Since *Tobacco Road*: An Interview with Erskine Caldwell." *Southwest Review* 69, no. 1: 33–47.

Kevles, Daniel J. 1985. *In the Name of Eugenics: Genetics and the Uses of Human Heredity*. New York: Knopf.

Klevar, Harvey L. 1993. *Erskine Caldwell: A Biography*. Knoxville: University of Tennessee Press.

MacDonald, Scott. 1979. "Enough Good Reasons for Reading, Studying, and Teaching Erskine Caldwell." *Pembroke Magazine* 11: 7–18.

Miller, Dan B. 1995. *Erskine Caldwell: The Journey from Tobacco Road*. New York: Knopf.

Nordan, Lewis. 1995. Foreword to *Tobacco Road*, by Erskine Caldwell. Athens: University of Georgia Press.

Pollock, Channing. 1936. "The Survival of the Unfittest." *Reader's Digest* (November): 37–40.

Reiss, Benjamin. 1999. "P. T. Barnum, Joice Heth, and Antebellum Spectacles of Race." *American Quarterly* 51, no. 1: 78–107.

Todorov, Tzvetan. 1984. *Mikhail Bakhtin: The Dialogic Principle*. Translated by Wlad Godzich. Minneapolis: University of Minnesota Press.

Wiggam, Albert Edward. 1927. "New Styles in Ancestors." *World's Work* (December): 142–50.

Reproducing the Working Class: Tillie Olsen, Margaret Sanger, and American Eugenics

Claire M. Roche

Introduction

CAN WRITERS REPRESENT PEOPLE WITHOUT OBJECTIFYING THEM? Who has the right to represent a minority or marginalized group of people? What happens as a result of the representation, or what work does the representation do? Who speaks for whom, and who has the power to speak? These questions make clear what W. J. T. Mitchell suggests: "representation, even purely 'aesthetic' representation of fictional persons and events, can never be completely divorced from political and ideological questions" (1995, 15). In other words, representations are always already ideologically and politically informed, even if we are not aware of it, and despite our intentions. Further, to the extent that any writer creates representations within linguistic and generic forms recognized by a particular ideological position, that writer runs the risk of reproducing the stereotypical and essentialist aspects of the thing that she is trying to represent. Tillie Olsen's *Yonnondio* is one example of a work of fiction that represents the working class and, in that representation, reproduces the problematic ideologies and notions of the working class that it works to subvert.

Margaret Sanger's *Birth Control Review*, like *Yonnondio*, presents often troubling and ideologically-informed representations of the working class, particularly as they pertain to human reproduction and the American eugenics movement. Further, reading excerpts from Sanger's publication against passages from Olsen's *Yonnondio* makes clear the difficulties of representing an ideologically loaded and overdetermined category of subjectivity such as the "working class." While both women sought what they considered progressive social change, Sanger's political and ideological agendas were significantly different from Olsen's. Sanger used ideologies in intentional ways in her work, while in Olsen's case, ideology "uses" her work regardless of her intention.

Ideology and representation are, of course, closely related concepts. As Kavanagh argues, "ideology is less tenacious as a 'set of ideas' than as a system of representations, perceptions, and images that precisely encourages men and women to 'see' their specific place in a historically peculiar social formulation as inevitable, natural, a necessary function of the 'real' itself" (1995, 310). That said, what is it we "see" when we read *Yonnondio*, either by itself or with Sanger's *Birth Control Review*? What we see is another in an endless series of representations, including those furthered by eugenicists, of the working class as "other," as "less than," as socially undesirable. Although Olsen was working in conjunction with a "progressive social struggle," she faced this struggle of representation in the dominant ideological realities of her time. The ideological apparatuses against which she worked historically were 1930s views on class relations, labor, production and reproduction, eugenics, and Social Darwinism. Whatever her intentions, she could not efface the power of the dominant ideological systems that necessitated that the novel be written in the first place.

The ideological systems Olsen faced resulted in part from the reform efforts of the Progressive Era, including Sanger's. The period from 1890 to 1920 is one in which the American romance with progress was at its peak. And while progressivism has typically been associated with liberals or the left (read: those with an assumed altruistic interest in social reform), historians don't agree on what defines the legacy of the Progressive Era. All manner of things have been done in the name of progress, as we know. We can't immediately assume that reform and progress, then or now, are always good for those people in whose name they are sought. In other words, we can't fit all eugenicists into one political or ideological category. Neither can we assume that a particular political position or identification was always tied to particular motives or agendas. As historian Colin Gordon argues, "Progressive political thought, after all, shared at least one thing with modern liberalism: it was irretrievably elitist in outlook and often fundamentally undemocratic in practice" (1995, 672).[1] I am not suggesting that Olsen or Sanger were "fundamentally undemocratic." In fact, it is clear that their careers as activists were devoted to democratization in some form. However, their careers did not take place in a vacuum. That is, how democratization signifies is also historically and culturally determined. As Gordon suggests, the Progressive Era was characterized by a degree of arrogance that "emerged in both Progressive efforts to shape and cultivate public opinion, and in the logic of Progressive political reform—which tackled not only the corruption of machine politics but also the threat of class politics and mass democracy in local and national settings" (672–73). The America about which Olsen wrote in *Yonnondio* was one that was subject to the legacies of the Progressive Era. The America in which Sanger worked was no less complex and contradictory.

Birth Control Review

Margaret Sanger founded the birth control movement in America, numerous birth control "leagues," and *Birth Control Review*, a magazine that announced itself as being "Dedicated to the Cause of Voluntary Motherhood." *Birth Control Review* was published from 1917 until the early 1940s, and with the exception of the last several years of its publication, Sanger was either contributor to and editor of the publication or involved in some other way. While Sanger wrote several books and was involved in other publications, *Birth Control Review* was her earliest and perhaps most interesting publication outlet, one that contains numerous representations of the working class. And while I am most interested in and intrigued by "Letters from Mothers," a regular feature of the magazine, an overview of the history of the publication, as well as Sanger's career and the movement she founded, is necessary.

Sanger occupies an unusual place in American history and the history of women. There are few other people who worked as diligently and tirelessly as Sanger did to effect change that would be valuable to women. The current nationwide network of Planned Parenthood chapters and clinics is genealogically connected to the first women's reproductive health clinic opened in this country, by Sanger, in Brooklyn, New York. Because of Sanger's lifelong efforts, women today have access to a variety of birth control methods and devices. Yet, however valuable Sanger's contributions, her career, her rhetoric, and her ideologies are not without problems.

While much of the scholarship on Sanger and her career is limited in so far as it does not (refuses to?) recognize the contradictions in both her politics and the movement she engineered, there are nonetheless several valuable sources. Ellen Chesler, author of *Woman of Valor: Margaret Sanger and the Birth Control Movement in America*, is one of the few scholars who recognizes the contradictions between Sanger's rhetoric and practice, between her ideology and her goals. Sanger became involved in the birth control movement for a variety of reasons, not the least of which was that she saw firsthand, as a nurse working on the Lower East Side of New York, the devastating effects on women and their families of the combination of poverty and too frequent child bearing. In the earlier years of the movement, she would also become acquainted and then intimately involved with Havelock Ellis, the influential sociologist, "sexologist," and eugenicist. Ellis's position on eugenics is summed up by his own statement that appeared in the "Havelock Ellis Number" of *Birth Control Review* February 1919 issue: "We desire no parents who are not both competent and willing parents. Only such parents are fit to father and mother a future race worthy to rule the world" (8). Ellis would publish in *Birth Control Review* frequently, and would shape Sanger's ideas in significant ways. As Chesler says, "Ellis always considered himself both a eugenicist and a socialist and convinced Margaret of the coherence of this viewpoint. Ellis made his most important contribution to the eugenics doctrine, at least from the standpoint of Margaret's interest, when he assigned women to act as its chief enforcers. . . . Increased sex expression and wider use of birth control were thus significant tools in the eugenics program and accordingly, he condemned eugenicists who refused to endorse birth control because they wanted more children for the better classes" (1992, 123). It is this question of class as it is tied to the history of eugenics that renders Sanger's work and legacy most interesting, important, *and* troublesome.

Sanger, at turns, strongly adopted or shied away from eugenics as a means to further her cause in making access to birth control a possibility for all women. Apparently, she was either unable or unwilling to recognize the complicated nature of this association. Sanger marketed herself and her movement as, at times, liberal, leftist, Socialist, and progressive, while she supported and made use of the increasingly elitist views of the eugenicists working in the United States and England. Sanger did not, within the pages of *Birth Control Review*, write her own opinions about eugenics or the fitness of the working class as parents. Rather, she made clear her views in these areas through other publications, such as *The Pivot of Civilization* and *Woman Rebel*. In addition, in her book, *Women and the New Race*, Sanger made it clear that birth control was most important with respect to controlling the numbers of "unfit" in the population: "Birth Control itself is nothing more or less than the facilitation of the process of weeding out the unfit, of preventing the birth of defectives or of those who will become defective" (1920, 105). And while Sanger would, over the years, redefine what she meant when she invoked the class of the "unfit," ". . . she increasingly saw feeblemindedness, the bogey of all hereditarians, as antecedent to poverty and social organization in the genesis of social problems" (D. Kennedy 1970, 115).

Sanger's working relationship with eugenicists would play itself out in the pages of *Birth Control Review*, from its inception through the 1930s, as article after article appeared in the debate about the role of birth control in eugenics programs. In other words, *Birth Control Review* reflected the scientific and social thinking in its historical moment. As Chesler argues, in the 1920s "eugenics became a popular craze in [America]—promoted in newspapers and magazines as a kind of secular religion. A national advocacy organization . . . was founded . . . to foster broader public understanding of eugenic principles through such public relations gimmickry as . . . 'fittest family' contests at state fairs" (1992, 215). As represented by the sweeping claims published in *Birth Control Review*, the influence of eugenics cannot be overlooked: "The Eugenic touch-stone is the final and infallible test of all ethics and all politics" (R. Kennedy 1921, 17). Integral to this test and the eugenics platform in the 1920s and later was a belief that sterilization was a viable method of birth control, and a method that could be legislated and enforced on whole groups of the population to include those deemed "feebleminded." Despite the significant ethical and moral implications of sterilization programs, Sanger would continue to associate her movement with that of the eugenicists, in part to defend against attacks from religious institutions, especially the Catholic Church (Chesler 1992, 216). If politics makes for strange bedfellows, so too did the advocacy for birth control.

Despite these apparent ideological and moral quagmires, Sanger believed that so long as she continued to advocate for "reproductive choice" for women, she was "working in accord with the universal law of evolution" (quoted in Douglas 1970, 130). The prominent eugenicists publishing in her magazine would concur. Paul Popenoe, a highly influential eugenicist who was one of the first proponents and scholars of state-sponsored sterilization programs, and editor of the *Journal of Heredity*, contributed a typically hyperbolic and polemic article to the third issue of *Birth Control Review*. While Popenoe argued that birth control was not eugenic in so far as it might reduce the numbers of children born to parents of good stock, that is, birth control was dysgenic, he went on to argue that the "less capable" part of the population should not only practice birth control, but should be forcibly sterilized. He argued for sterilization, in part, because birth control was not enough to "affect the reckless and improvident, those who procreate while drunk—those, in short, whose children the race would be better off without" (1917, 6). There were eugenicists of Popenoe's stripe, clearly social conservatives and elitists. There were also advocates, such as Sanger, whom we might call liberal, who had in mind the improvement of women's reproductive health and improved quality of life for their children, at least initially. What, then, to make of the birth control advocate "in bed with" the eugenicists like Popenoe?

Sanger herself identified, in a later issue of her magazine, that "the campaign for Birth Control is not merely of eugenic value, but is practically identical in ideal with the final aims of Eugenics" (1921, 5). On the first page of the November 1921 issue of her magazine appeared an epigraph which read: "Birth Control: To Create a Race of Thoroughbreds" (3). Few of Sanger's contributors supported this goal as ardently as Anna Blount, a medical doctor and Chairman of Eugenics Education Society of Chicago. Drawing heavily on the rhetoric of Social Darwinism, Blount argued that the working class, particularly the noncontributing poor, should not be allowed to "breed" indiscriminately. She sided with the eugenicist, a "true radical," who worked to "purge the world of imbeciles, epileptics and the insane by ceasing to breed them. With no harm to any human individual, he would eliminate in time those of surpassing moral and physical ugliness" (1918a, 7). In an article entitled, "Large Families and Human Waste," Blount would increase the stakes for the working class, those of the "unlovely face": "There they are, a motley group, from the gay, light-hearted moron, who cannot make an intelligent plan, even to do mischief; to the doddering idiot, the crafty paranoiac, the wretched epileptic, the moral imbecile, the chronic criminal with hereditary taint, and even the village ne'er-do-weel. What do they cost us, in wealth, in labor and in misery? *They must be eliminated.* Eugenics makes birth control imperative. Defectives may be segregated, they may be sterilized, and the brighter ones of them may learn methods of contraception. Their marriages should be forbidden, as an expression of the public will that their children are not wanted. But whatever the means this stream of human waste must be deflected from the melting-pot" (1918b, 4) (emphasis added).

What, exactly, does a phrase like, "the village ne'er do weel" mean? Whom does it designate or signify? In a passage that eerily foreshadows the horrors of the Nazi "final solution," Blount, like Sanger, targeted a large portion of the working class as a social evil. The worker who was moral and intelligent wasn't the problem (I wonder, would Blount or those who shared her views have argued that they were hard-pressed to find these workers?), but the members of the working class who somehow "cost" the middle and upper classes something—money, trouble, discomfort—should, as a minimum, be practicing birth control, or be eliminated altogether.

So while Sanger, on the one hand, worked for the right for women to have reproductive choices, she also worked to dictate which choice women might make or which women had to choose. Nowhere else in her rhetoric is this made clearer than in the contradiction between the representations of the working class found in the pages of *Birth Control Review* and the letters submitted to Sanger and her publication by working-class women. It is as though working-class women were invited to write to Sanger or the magazine simply in order to further objectify themselves through original textual proof that eugenics programs were needed. It is no coincidence that the letters are from poor and/or working-class women. The histories of both the birth control movement and medical practice in this country make clear that middle- and upper-class women often had access to information while poor and working-class women did not. Further, to the extent that Sanger courted eugenicists in order to garner their support for her movement, she would not have printed letters from middle- or upper-class women who supported birth control. They would have been unnecessary if not counterproductive in so far as they might be considered dysgenic.

Following are several examples of letters received from women who either knew who Sanger was and/or who were readers of *Birth Control Review*. While it seems to me that the letters speak for themselves, I think it is important to point out that the letters, read with or against the rhetoric of the "professional" contributors, highlight the gaps between Sanger's rhetoric and her practice, as well as the nature of representations of the working class that go without interrogation of any kind.

From the inaugural issue (February 1917) of *Birth Control Review*:

> Mrs. Margaret Sanger:
>
> I read about you in the paper. I am a poor man's wife. We have nothing but our little children. We have had six children and we are not able to feed and clothe them. I am in very poor health, and I think it is a sin for me to have to raise any more children. You have a pamphlet on birth control. Can you let me have one?
>
> We are very poor and money is scarce with us, but I am sure you have some little idea how the poor has to live.
>
> Please let me hear from you.
>
> From a sister.
>
> Mrs. A. J.________(5)

From the June 1918 issue:

> Letter No. 17
>
> I have read some of your books and have heard people speak of you, and now I am going to ask you for a little advice or help, please. I am just 22. I have been married 6 years, I have three children, the oldest is five years old and my baby is 11 months old. My husband has tuberculosis and has been to the state sanatorium and as yet can keep working. My husband's parents are dead and we had to take their children three of them not large enough to work for themselves, and the other to pay board. There is five of them and five of us with baby making ten in family and I have to do all the cooking, house-work and sewing for all and my baby to attend to also. I am telling you this so you won't think I am foolish for writing you, still I think I go crazy thinking about everything. Now, what I want to tell you is I have missed my monthly sickness once and I am afraid I am pregnant as I never miss unless I am. I don't know how it happened as I've been careful since my husband has been sick. I have taken laxatives and quinine and it has failed. I have been to my family doctor, and he knowing how it is with me refused to help me. I would pay you whatever you charge and keep quiet about it, and be much obliged to you. If you can help me any or can't please answer this so I won't be writing and bothering you. Sincerely, Mrs. C. M. C. (13)

In the November 1918 issue, the editors (possibly including Sanger) changed the layout of the letters page to include tabloid-like headlines. The page itself is headed with: "Verboten! Verboten! Verboten!", referring to one physician's response to a patient's request for information about how not to get pregnant.

Husband Constantly Drunk.

Dear Madam:—
Owing to the fact that I am compelled to go to work and would think it a great favor if you could kindly send me your advice on birth control. My reasons are that *I have a husband who is constantly drunk and in order to give my children enough to eat I must go to work*. I am 28 years old, have four children, youngest 2 years. Kindly oblige as I am ready to do anything in the world for my children and feel it would be a sin to bring more into the world. I think you are doing the finest work in the city and wish you luck. (10) (emphasis in original)

Ten Dollars a Week for Five.

Dear Madam:—
I ask a great favor of you to send me one of your pamphlets as I am married six years last July and have had five children, three living, 2 dead, the youngest 1 ½ years. My husband does not earn more than ten dollars a week and find I have all to do to get along. If it were not for my folks I would go hungry many a time. (10)

In publishing these letters from mothers, Sanger was publishing a very particular representation of what are clearly working-class women, and she did so with a specific, if not contradictory, goal—to further the birth control movement with all its ideological baggage and, sometimes, eugenic elitism, as it appeared in the form of articles by eugenicists condemning or pathologizing these very women.

Clearly, little distinguishes the representations captured in these letters from the representations of working-class mothers as rendered by Tillie Olsen in the creation of her character Anna Holbrook. The only significant difference, when all is said and done, is genre. Ideologically, the representations in both have much the same effect. That Olsen was herself a working-class mother does not foreclose the possibility of the reproduction of stereotypes or misreadings of the working class in her writing.

Yonnondio

Tillie Olsen, much like Margaret Sanger, is something of an icon. She, like Sanger, represents the possibilities of women's involvement and leadership in social change. Olsen is considered by many critics to be a sort of Ur-feminist; she was writing what might be considered "modern" feminist literature long before anyone else. Her influential *Silences*, published in 1978, captures all of the frustrations and powerful implications of being a woman who writes. That she was also active in 1930s working-class politics and agitation contributes to her reputation as a significant and influential American woman writer. As Mickey Pearlman suggests, "Olsen is known and admired much more because of what she represents than because of what she has written" (1991, ix). But her work is unquestionably important to any understanding of the modern women's movement, particularly in literary studies. As a result, many of her critics seem to begin their work with a sort of blindness.

Regardless of the exuberance of some of Olsen's critics, or perhaps because of it, other critics, such as Pearlman and Werlock (1991), Coiner (1996), and Frye (1995), recognize the ideological traps that Olsen set for herself in attempting to represent the working class in America as ultimately human and dignified while using cultural stereotypes to do so. These critics also recognize the difficulties in representing motherhood, especially if the mother is working class. While Olsen worked throughout her career to subvert the stereotype of working-class mothers as "other," in calling them "essential angels," she also implied that working-class mothers wanted simply to inhabit the institution of middle-class motherhood in all its ideologically driven glory.

In *Yonnondio* we find all the difficulties of life for people of the working class, including motherhood. The Holbrooks are a poor, working-class family who move from one part of the Midwest to another in search of steady employment, safety, and enough of anything—enough money, enough food, enough self-respect. As the novel opens, the Holbrooks are living in a mining town in Wyoming. Anna and Jim Holbrook already have more children than they can feed. The oldest, Mazie, asks Anna what there is to eat. Anna's reply is, "Coffee," which is, of course, absent from any nutritionist's food charts (Olsen 1974, 3). Immediately thereafter, Mazie asks her mother what an "edication" is. Anna's reply represents the working-class perception of the middle class: "An edjication? . . . An edjication is what you kids are going to get. It means your hands stay white and you read books and work in an office. Now, get the kids and scat. But don't go too far, or I'll knock your block off" (3). Anna is a poor mother with too many children, too little time and patience, and a short temper, not unlike many of the mothers who wrote letters to Sanger's magazine.

There are numerous didactic and relatively heavy-handed moments in *Yonnondio*, on which several of Olsen's critics comment. Coiner has the following to say about Olsen's work in this novel: "In the tradition of muckraking, of social journalism, and reportage, Olsen's writing during [the 1930s] . . . was intended to spark political resistance. At the same time, however, its didacticism . . . actually limits involvement by undercutting the reader's role as an active producer of meaning" (154–55). It is at these moments, when the reader is excluded from the construction of meaning, that the representations in *Yonnondio* become a source of concern. One of these moments comes to pass when Sheen McEvoy tries to sacrifice Mazie to the mine. Crazy McEvoy thinks that the mine demands a sacrifice in exchange for the safety of the men who work in the mine. McEvoy just happens to encounter Mazie while he is well into a hallucinatory episode. He says, "The mine is calling for her baby," then finds Mazie, and picks her up to throw her into the mouth, or "Iron Throat," of the mine. He says, "Give her [the mine] a sweet baby, and she'll want no more" (12). However didactic this passage may be, it is also quite complex. Clearly, the message for the reader is that the working class must sacrifice a great deal, even their families, to continue working, to buy their safety. But it is a girl who will be thrown to the mine, a woman who will make the sacrifice so that the men can keep working and ensure the future. As Pearlman suggests, "the answer to oppression, the hope of the future, is embodied in young women who will fight to improve life for their children" (22). This answer appears throughout *Yonnondio*, in a variety of manifestations.

The Holbrooks leave Wyoming for an unsuccessful attempt at sharecropping in South Dakota and then move to Kansas City where Jim Holbrook works in the sewers of the city and then the meat packing houses. As Jim moves his family from one bad situation to one that is worse, as the work he does becomes more brutal and degrading, his behavior and treatment of his wife and children become more brutal as well. After Anna has become pregnant yet again, and then miscarried, Jim rapes her, and Mazie hears the whole thing through the too-thin walls of their tenement apartment. Anna says to him, "Dont, Jim, dont. It hurts too much. No, Jim, no." Jim, however, pays her no mind, and says, "Cant screw my own wife. Expect me to go to a whore? Hold still" (75). As the story progresses, Jim drinks more frequently and leaves his family for extended periods, only to return to get Anna pregnant again. To whatever extent Olsen might have been trying to represent the brutality of the capitalist system that put Jim in such a position, she also gives us the brutal working class, the "village ne'er-do-weel" who appears in Anna Blount's representation of the working class. As Abigail Martin argues, "Olsen does the working class no good service when she shows only the seamy side of life for the American working class" (1984, 36).[2] Like the husbands of Sanger's mothers, "Jim Holbrook, [is] a 'good' and well-intentioned man, [who] refuses to accept full responsibility for his actions and wavers between the terrible non-choices of abandonment, alcoholism, and abusive behavior" (Staub 1988, 135). Pearlman and Werlock also argue that "Olsen's point is that Jim blames Anna and her 'goddamn brats' for his failure, and she blames him. Olsen rightfully 'blames the system' " (1991, 44). Yet Olsen, simply by virtue of representing the system that oppresses the Holbrooks, runs the risk of reproducing that very system and affirming the eugenic claims about the unfitness of the working class. Joanne Frye also has a point to offer on this issue: "any attempt to make women's lives or working-class lives central to a narrative form . . . is at risk of being reassimilated into patriarchal constructions and prevailing definitions of the meanings of those lives to the forms and linguistic expressions that are already prevalent in society. The experiences of such people . . . cannot be simply conveyed in the language or literary forms of the dominant culture" (1995, 11). Nowhere are these risks more apparent, and nowhere are the similarities in representation to Sanger's work more apparent than in the scene in which a doctor is called in to examine the youngest of the Holbrooks, Bess, who is ill, and to attend to Anna following her miscarriage.

Olsen clearly makes an effort in this scene to represent the doctor as insensitive and uncaring, and he arguably is. But he is also a patriarchal medical professional and, to some extent, representative of some of Olsen's readers. And given his cultural position, he has a public voice, while the Holbrooks go unheard. Jim has gone for the doctor, and in his absence, the children wait anxiously. Upon the return of the men, the doctor says, presumably to Jim, " 'Miscarriage. You didn't know she was pregnant—again?' " (Olsen 1974, 77). Then begins a series of lines of dialogue spoken by the doctor, followed by parenthetical asides:

> "How old's the baby?" (Damn fools, they ought to sterilize the whole lot of them after the second kid.)

> "Four months, mm. You remember how long your wife's been feeling sick? Of course not." (These animals never notice but when they're hungry or want a drink or a woman.)
>
> "Hmmm. Yes." She took the ergot down quietly, but moaned at the hypo. "So you had intercourse before, it wasn't only the fall." (Pigsty, the way these people live.) "And she's been nursing all along? We'll have a look at the baby." (Rickets, thrush, dehydrated; don't blame it trying to die.) (77)

The good doctor in *Yonnondio* has the same contempt for the poor working class as Anna Blount, M.D. Of course, it is clear that we are meant to disapprove of the doctor in this scene, but as W. T. Mitchell asks of Robert Browning's Duke in "My Last Duchess," "what form does this disapproval take?" (1995, 19). After all, the eugenics movement was still very much a reality in 1930s America, and sterilization was of central concern. To the October 1932 issue of *Birth Control Review*, C. O. McCormick, M.D. contributed an article entitled, "Eugenic Sterilization." In it McCormick argued for a solution to the "serious [truth] that our growing population is being increasingly maintained by the moron group" (242). He suggests that a "ready and feasible plan would be selective and eugenic sterilization, in that it is a direct means of relieving present and future society of social and financial burdens by eliminating mental and physical defectives and thus insuring progressive improvement of the race" (242). It seems the doctor in *Yonnondio* would like to relieve society of the Holbrooks, and the Holbrooks would very likely qualify for McCormick's "compulsory group" for sterilization, which included "the recovered and unsegregated insane, feeble-minded, epileptics, habitual criminals, and chronic paupers" (242).

Much of the difficulty that the Holbrooks encounter, as a working-class family, results, of course, from repeated and unwanted pregnancies. With each pregnancy, there is an inversely proportional likelihood that the Holbrooks will succeed in surviving at all, much less in ever moving beyond their working-class status. And with each pregnancy, Olsen increases the potential that her representation of the working-class family will mimic that available in Sanger's publications. Of course, Olsen also works to explode the myth of the beauty of motherhood—women like Anna Holbrook have neither the time, money, nor energy to revel in their children in ways that follow the traditions of the Virgin Mother or any of the views of motherhood available in the abundance of women's periodicals available in America. But as Pearlman and Werlock point out, "Almost twenty years [after the publication of *Yonnondio*], this message has not penetrated the minds of some critics" (1991, 43). They make a good point, and this may again be a question of the inherently problematic nature of representation: how to represent motherhood in a way that *won't* reproduce existing ideological apparatuses that define motherhood as an institution to which only particular women have access; and in the 1930s, how to represent working-class motherhood without falling into the trap of eugenics. How do we, in fact, represent working-class women without reinscribing women's bodies as the site of the eugenics battleground?

Yonnondio may be considered a realistic novel, even if there are elements that do not make attempts at realism. Ideological apparatuses are fully embroiled in realism, as they are in other forms or conventions of literature. Frye, drawing on the work of Terry Eagleton, argues that, "'ideology, presenting itself as life, is the basis for the privileged epistemological standpoint of realism.' The traditional narrative voice of realist fiction—fiction that claims to represent people's actual lives—must apparently be grounded in the predominant values of that fiction's culture. This, then, is the problem that haunts Olsen's fiction" (Frye 1995, 12). Olsen departs at times from the "traditional narrative voice," as she attempts to write in the style of Bahktin's "polyphonic" form. However, as long as she attempts to render working-class experience in a realistic form, she cannot get beyond that form and its attendant ideological positions and implications, which include the reproduction of the working class as "submerged," in Sanger's terms, or "unfit" in eugenic terms.

Few critics of *Yonnondio* include in their discussion the "fragments" that Olsen appended to the novel after it was rediscovered in the 1970s. But here, again, is more of the same—more ways in which Olsen's representations of the working class can be read as potential conflations with predominant cultural views of the working class: Anna gets pregnant again and attempts to abort her own baby, only to end up in the care of nuns; Jim is in and out of work, in and out of the family scene, is "falsely charged with murder" (1974, 136), and ignores Anna's request that he use a rubber during sex in order to prevent the pregnancy that she will later attempt to abort; Mazie has gone to work packaging nuts and dried fruits for a merchant who molests her as payment for nuts she took home. All of this is in stark contrast to the end of the 1930s version of *Yonnondio*, which critics argue is hopeful.

The original version ends with Bess, still just a baby, banging a jar lid on a table. Olsen renders the scene as a sort of evolutionary zeal that Bess feels and that ends the story on the hopeful note that the Holbrooks, and by extension, the working class, will evolve, succeed, and survive: "Bess who has been fingering a fruit-jar lid—absently, heedlessly drops it—aimlessly groping across the table, reclaims it again. Lightning in her brain. She releases, grabs, releases, grabs. I can do. Bang! I did that. I can do. I! A look of neanderthal concentration is on her face. That noise! In triumphant, astounded joy she clashes the lid down. Bang, slam, whack. Release, grab, slam, bang, bang. Centuries of human drive work in her; human ecstasy of achievement; satisfaction deeper and more fundamental than sex. *I can do, I use my powers; I! I!* Wilder, madder, happier the bangs. The fetid fevered air rings with Anna's, Mazie's, Ben's laughter; Bess's toothless, triumphant crow. Heat misery, rash misery transcended" (132).

As a reader, I find it difficult not to get caught up in Olsen's enthusiasm. Her writing here is powerful and engaging, if not convincing. It is this nearly final scene on which so many critics focus their arguments that Olsen's work is about the triumph and transcendence of the working class. But Olsen wrote, later, on the very next page, "Reader, it was not to have ended here . . ." (133). The ending of the final 1970s fragment takes place in a ward of some sort, after Anna has attempted to abort her child using a pair of scissors. The last sentence reads, "At last she forgot her children, forgot herself, was only pain, pain, and in the core a small child shrieking Momma momma momma for a mother eighteen years dead" (151). Here Bess's achievement in the evolution of motor skills is forgotten and irrelevant. Anna Holbrook, the working-class mother, is "only pain, pain." This representation, perhaps better than any other, offers a fictional version of the paradox Olsen faced as a leftist writer whose work can be read as having a conservative agenda. As Frye points out, "the capacity of language for reinforcing conformity does make it risky to pursue 'experience' as a basis for literary form, since the politics of experience is inevitably (or at least may become) a conservative politics for it cannot help but conserve traditional ideological constructs which are not recognized as such but are taken for the 'real'" (Frye 1995, 11).

The work that Olsen's novel "does," the ideological implications of her representation of the working class, is not something for which she is to be blamed. If nothing else, the reading I've presented highlights the ideological trap of representing anything in language. Olsen, unlike so many of her contemporaries and followers, actually tried to experiment with language, to alter form in ways that might result in a representation outside a dominant ideology. Nonetheless, the contributions of both Sanger and Olsen to the modern women's movement and to women's reproductive and political choices are significant, and we continue to reap the rewards of those contributions today. But their representations of the working class, specifically working-class women, are not without cultural and ideological implications. More importantly, and more interesting, however, are the questions that the work of both women lead us to ask: Can we represent experience without reproducing dominant ideologies? What forms are available to us to do this, and what forms have yet to be created? What happens to our understanding of "the working class" if we avoid conflating it with reproduction, in whatever context we might invoke it? Can reproduction be theorized as separate from class? What possibilities does the work of either woman offer us on which we can build and experiment? As these questions suggest, we must continue to interrogate theories of representation and the ideological implications of those theories.

Notes

1. Gordon writes primarily, though not exclusively, of political reform in terms of cleaning up political corruption. Social reform is inseparable from political reform, particularly to the extent that social reforms have political ramifications.
2. Martin's remark highlights yet another challenge inherent in any attempt at representation. A writer who puts too fine a face on the poor would be critiqued for creating a representation that was too idealistic. Negotiating this gap—between a representation that is somehow too "real" and a representation that is too "unreal"—may be one of the greatest difficulties writers face, including Tillie Olsen.

References

A. J., Mrs. 1917. "Letters." *Birth Control Review* 1 (February): 5.

Blount, Anna E., M.D. 1918a. "Eugenics in Relation to Birth Control." *Birth Control Review* 2 (June): 7, 15.

———. 1918b. "Large Families and Human Waste." *Birth Control Review* 5 (September): 3–4.

C. M. C., Mrs. 1918. "Letter No. 17." *Birth Control Review* 2 (June): 13.

Chesler, Ellen. 1992. *Woman of Valor: Margaret Sanger and the Birth Control Movement in America*. New York: Anchor.

Coiner, Constance. 1996. "Literature of Resistance: The Intersection of Feminism and the Communist Left in Meridel Le Sueur and Tillie Olsen." In *Radical Revisions: Rereading 1930s Culture*, edited by Bill Mullen and Sherry Lee Linkon. Urbana: University of Illinois Press.

Douglas, Emily Taft. 1970. *Margaret Sanger: Pioneer of the Future*. New York: Holt, Rinehart, and Winston.

Ellis, Havelock. 1919. "Birth Control, Morality and Eugenics." *Birth Control Review* 3 (February): 8.

Frye, Joanne S. 1995. *Tillie Olsen: A Study of the Short Fiction*. New York: Twayne.

Gordon, Colin. 1995. "Still Searching for Progressivism." *Reviews in American History* 23: 669–74.

"Husband Constantly Drunk." 1918. *Birth Control Review* 2 (November): 10.

Kavanagh, James H. 1995. "Ideology." In *Critical Terms for Literary Study*, edited by Frank Lentricchia and Thomas McLaughlin. Chicago, IL.: University of Chicago Press.

Kennedy, David H. 1970. *Birth Control in America: The Career of Margaret Sanger*. New Haven, Conn.: Yale University Press.

Kennedy, Robert H. 1921. "The Eugenic Conscience." *Birth Control Review* 10 (June): 17.

Martin, Abigail. 1984. *Tillie Olsen*. Boise, Idaho: Boise State University.

McCormick, C. O., M.D. 1932. "Eugenic Sterilization." *Birth Control Review* 16 (October): 241–42.

Mitchell, W. J. T. 1995. "Representation." In *Critical Terms for Literary Study*, ed. by Frank Lentricchia and Thomas McLaughlin. Chicago, IL.: University of Chicago Press.

Olsen, Tillie. 1974. *Yonnondio: From the Thirties*. New York: Delta.

———. 1978. *Silences*. New York: Delacorte.

Pearlman, Mickey, and Abby H. P. Werlock. 1991. *Tillie Olsen*. Boston: Twayne.

Popenoe, Paul. 1917. "Birth Control and Eugenics." *The Birth Control Review* 1 (February): 6.

Sanger, Margaret. 1918. "All Together-Now!" *Birth Control Review* 2 (October): 7.

———. 1920. *Woman and the New Race*. New York: Blue Ribbon Books.

———. 1921. "The Eugenic Value of Birth Control Propaganda." *Birth Control Review* 10 (October): 5.

———. 1922. *The Pivot of Civilization*. New York: Brentano's.

Staub, Michael. 1988. "The Struggle for 'Selfness' Through Speech in Olsen's *Yonnondio: From The Thirties*." *Studies in Short Fiction* 16: 131–39.

"Ten Dollars a Week for Five." 1918. *Birth Control Review* 2 (November): 10.

Contributors

DEBRA BEILKE, Associate Professor of English at Concordia University-St. Paul, teaches courses in writing, introductory literature, world literature, and American literature. She received her Ph.D. from the University of Wisconsin-Madison. Her research interests include southern literature, African literature, feminist theory, and women's writing. She has published articles on Zora Neale Hurston, Ellen Glasgow, Julia Peterkin, Frances Newman, and others. She is currently working on a book-length study of southern literary humor in the 1920s.

LOIS A. CUDDY received her Ph.D. from Brown University and is Professor Emerita, English and Women's Studies, at the University of Rhode Island. She is the author of *T. S. Eliot and the Poetics of Evolution* (BUP), co-editor of *Critical Essays on T. S. Eliot's "The Waste Land,"* and been assistant editor and reader for *ATQ*. She is the author of essays on poetry, fiction, and drama by various United States and British authors of the nineteenth and twentieth centuries.

CYNTHIA J. DAVIS is an Associate Professor of English at the University of South Carolina in Columbia, the author of *Bodily and Narrative Forms: The Influence of Medicine on American Literature, 1845–1915* (2000), and the coauthor of *Women Writers in the United States: A Timeline* (1996). She is working on a biography of Charlotte Perkins Gilman.

JUSTIN D. EDWARDS is Associate Professor of modern English literature at the University of Copenhagen. He is the author of *Exotic Journeys: Exploring the Erotics of U.S. Travel Literature* and *Gothic Passages: Racial Ambiguity and the American Gothic*, as well as co-editor of *American Modernism across the Arts*.

JULIA C. EHRHARDT is assistant professor of Honors and Women's Studies at the University of Oklahoma Honors College. Her work on Canfield constitutes part of a larger project on early twentieth-century women writers (including Zona Gale, Josephine Herbst, and Rose Wilder Lane) and the political agendas they advanced in their fiction.

SARAH C. HOLMES is Professor of English at the Naval Education and Training Center in Newport, Rhode Island. She is the editor of a volume of letters, *The Correspondence of Ezra Pound and Senator William Borah*. Her specialization is literature and culture of 1930s America.

LISA HOPKINS is a Reader in English at the School of Cultural Studies, Sheffield Hallam University, Collegiate Crescent Campus, United Kingdom. She is also editor of *Early Modern Literary Studies*: purl.oclc.org/emls/emlshome.html. She is working on a book-length study of popular fiction and evolution.

ANDREW LAWSON is a Senior Lecturer in American Literary and Cultural Studies at Staffordshire University, United Kingdom. He is currently working on a book on class identity and stylistic innovation in modern American writing. An article on Walt Whitman was drawn from this work and is forthcoming in *American Literature*.

JOHN NICKEL is a Ph.D. student in the Department of English and Comparative Literature at Columbia University. He is specializing in nineteenth- and twentieth-century United States literature. His articles have appeared in *The Concord Saunterer*, *Dictionary of Literary Biography*, *Essays in Arts and Sciences*, *The Eugene O'Neill Review*, *Textual Practice*, *Texas Studies in Literature and Language*, and *ATQ*.

PENNY L. RICHARDS, Research Scholar with UCLA's Center for the Study of Women, holds a Ph.D. in Education from the University of North Carolina, Chapel Hill, and did postdoctoral work in the history of special education at the University of California, Santa Barbara. She is an editor for the listservs *H-Education* and *H-Disability* and serves on the editorial and advisory boards for the forthcoming *Encyclopedia of Disability* (SAGE). Other research interests include topics on family, caregiving, and marginalized mothers.

CLAIRE M. ROCHE, a Ph.D. candidate in English at the University of Rhode Island, is working on an archival history of English

Studies, 1949–2001. Her primary interests are Rhetoric and Composition, American literature, and women's literacy. She has co-edited an anthology entitled *Making the Harm Visible: Global Sexual Exploitation of Women and Girls—Speaking Out and Providing Services*, co-authored an article in *Twentieth Century Rhetorics and Rhetoricians*, and published in *The Journal of Teaching Writing.*

ALEX VERNON teaches twentieth-century American literature and writing at Hendrix College in Conway, Arkansas. His scholarly specialities include war literature and literature and the environment. His co-authored 1999 *The Eyes of Orion: Five Tank Lieutenants in the Persian Gulf War* won an Army Historical Foundation Distinguished Book Award.

TAMSEN WOLFF is an Assistant Professor in the Department of English at Princeton University, where she specializes in modern and contemporary drama. She is currently working on a study of hereditary theory, performance, and early twentieth-century American drama.

ELIZABETH YUKINS is Assistant Professor of English literature at John Jay College of Criminal Justice, City University of New York. This essay is part of a larger study about the meaning of familial and social illegitimacy in twentieth-century American literature and culture. Her article on trauma and inheritance, entitled "Bastard Daughters and the Possession of History in *Corregidora* and *Paradise*," appears in *Signs: Journal of Women in Culture and Society.*

We thank Dr. Greg Clingham, Director of Bucknell University Press, for his generous support; Christine A. Retz and Mary Ann Hostettler at Associated University Presses for their creative work on the production and design of this volume; and Jamie Carr for the index.

Index